FIBER OPTIC SENSORS

FUNDAMENTALS AND APPLICATIONS

D. A. Krohn

INSTRUMENT SOCIETY OF AMERICA

FIBER OPTIC SENSORS — FUNDAMENTALS AND APPLICATIONS

Printed in the United States of America

We would like to thank the many suppliers who provided material for this book, and we regret any we may have inadvertently failed to credit for an illustration. On notification we shall insert a correction in any subsequent printings.

INSTRUMENT SOCIETY OF AMERICA
67 Alexander Drive
P.O. Box 12277
Research Triangle Park, NC 27709

Library of Congress Cataloging-in-Publication Data

Krohn, David A.
Fiber optic sensors.

Includes bibliographies and index.
1. Fiber optics. 2. Optical detectors.
I. Title.
TS1800.K76 1988 621.36′92 88-13049
ISBN 0-87664-997-5

Book design by Summit Technical Associates, Inc.

Dedication

To the employees of EOTec Corporation

Table of Contents

Introduction

1

Fiber optic sensors represent a technology base that can be applied to a multitude of sensing applications. The following are some characteristic advantages of fiber optics that make their use especially attractive for sensors:

- Nonelectrical
- Explosion-proof
- Often do not require contact
- Remotable
- Small size and weight
- Allow access into normally inaccessible areas
- Potentially easy to install
- Immune to radio frequency interference (RFI) and electro-magnetic interference (EMI)
- Solid-state reliability
- High accuracy
- Can be interfaced with data communication systems
- Secure data transmission
- Potentially resistant to ionizing radiation

Most physical properties can be sensed optically with fibers. Light intensity, displacement (position), temperature, pressure, rotation, sound, strain, magnetic field, electric field, radiation, flow, liquid level, chemical analysis, and vibration are just some of the phenomena that can be sensed. Table 1-1 lists some typical applications and the associated sensors required. All of those listed can be addressed fiber optically.

Table 1-1
Sensors for Various Applications

Applications	Sensor
Automated production lines (steel, paper, etc.)	Position, thickness, limit switch, break detection, velocity
Process control	Temperature, pressure, flow, chemical analysis
Automotive	Temperature, pressure, torque, gas detection, acceleration
Machine tool	Displacement, tool break detection
Avionic	Temperature, pressure, displacement, rotation, strain, liquid level
Heating, ventilation/air conditioning (HVAC)	Temperature, pressure, flow
Appliance	Temperature, pressure
Petrochemical	Flammable and toxic gases, leak detection, liquid level
Military	Sound, rotation, radiation, vibration, position, temperature, pressure, liquid level
Geophysical	Strain, magnetic field
Utility	Temperature, displacement, electric and magnetic field

Historically, applications utilizing fiber optics were oriented toward the very simple sensors, such as card readers for computers and outage indicators to determine if the lights in automobiles were blown out. Technology has taken fiber optic sensors to the other extreme. Ultrasophisticated sensors using interferometric techniques are in development. Compared to conventional displacement sensors, these devices have 4 to 5 orders of magnitude higher resolution. In general, such high sensitivities are not required for most applications except for military hydrophone and gyroscope use. Most industrial sensors fall somewhere in between these extremes.

Fiber optic sensors can be divided into two basic categories: phase-modulated and intensity-modulated sensors.[1-7] Intensity-modulated sensors generally are associated with displacement or some other physical perturbation that interacts with the fiber or a mechanical transducer attached to the fiber. The perturbation causes a change in received light intensity, which is a function of the phenonemon being measured. Phase-modulated sensors compare the phase of light in a sensing fiber to a reference fiber in a device known as an interferometer. Phase difference can be measured with extreme sensitivity. Phase-modulated sensors are much more accurate than intensity-modulated sensors and can be used over a much larger dynamic range. However, they are often much more expensive. For the most part, interferometers have found applications in military systems where cost is not necessarily a major consideration. On the other hand, intensity sensors are well suited for widespread industrial use.

In subsequent chapters, both sensor categories will be discussed in detail, but it is useful here to further define the basic sensing mechanisms.

Intensity-modulated sensors detect the amount of light that is a function of the perturbing environment, as shown in Figure 1-1.[8] The light loss can be associated with transmission, reflection, microbending, or other phenomena such as absorption, scattering, or fluorescence, which can be incorporated in the fiber or in a reflective or transmissive target. Intensity-modulated sensors normally require more light to function than do phase-modulated sensors; as a result, they use large core multimode fibers (or bundles of fibers). Transmission, reflection, and microbending sensor concepts are the most widely used. In essence, they are displacement sensors. Examples of how fiber optics can be used to detect position (displacement) and relate the movement to a physical property include the movement of a diaphragm for pressure or the movement of a bimetallic element for temperature.

Phase-modulated sensors use interferometric techniques to detect pressure, rotation, and magnetic field, the former two applications being the most widely developed. Figure 1-2[8] shows a schematic representation of a Mach-Zehnder interferometer. The laser light source has its outgoing beam split such that light travels in the reference single-mode fiber and the sensing fiber, which is exposed to the perturbing environment. If the light in the sensing fiber and the light in the reference fiber are exactly in phase upon recombining, they constructively interfere with an increase in light intensity. If they are out of phase, destructive interference occurs and the received light intensity is lower. Such devices experience a phase shift if the sensing fiber under the influence of the perturbing environment has a length or refractive index change, or both.

The basic sensor types will be discussed in detail in further chapters as well as how these concepts can be applied to a wide variety of applications. Chapter 2 describes the fundamental principles of fiber optics. While it is not necessary to read the chapter to understand the sensing concepts, it is helpful and provides a background.

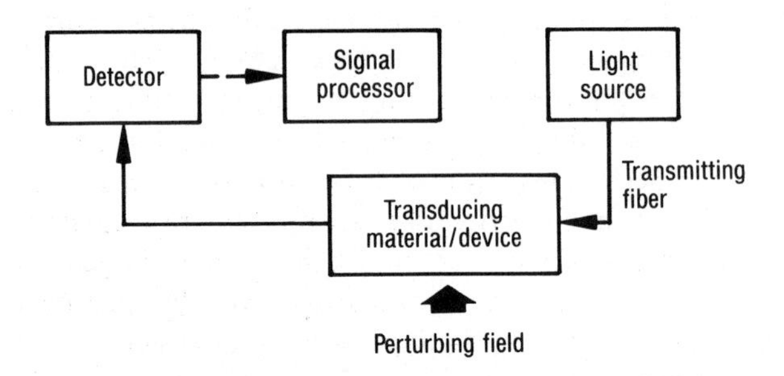

Figure 1-1
Intensity Sensor

(reprinted by permission from SRI International)

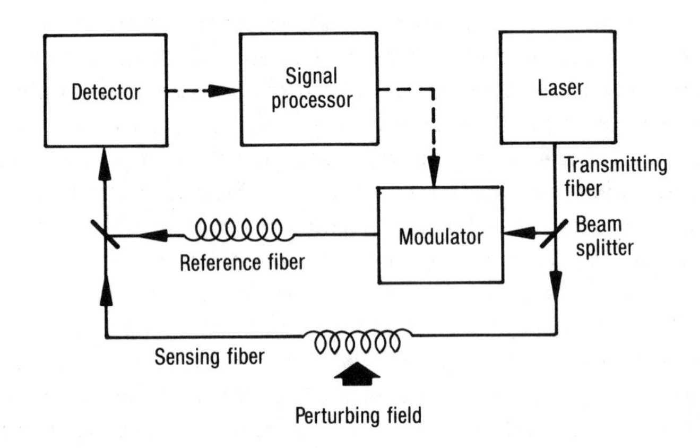

Figure 1-2
Phase Sensor

(reprinted by permission from SRI International)

REFERENCES

1. Davis, C. M., et al., 1982, *Fiber Optic Sensor Technology Handbook*, Dynamic Systems, Reston, Virginia.

2. ME Staff, May 1984, "The Exciting Promise of Fiber-Optic Sensors," *Mechanical Engineering*, pp 60–65.

3. Krohn, D. A., October 1982, "Fiber Optic Sensors in Industrial Applications: An Overview," *Proceedings of the ISA International Conference — Philadelphia*, Vol. 37, Part 3, pp 1673–1684.

4. Krohn, D. A., October 1983, "Fiber Optic Sensors in Industrial Applications: An Update," *Proceedings of the ISA International Conference — Houston*, Vol. 38, Part 2, pp 877–890.

5. Hecht, J., July–August 1983, "Fiber Optics Turns to Sensing," *High Technology*, pp 49–56.

6. McMahon, D. H., Nelson, A. R., and Spillman, W. B., December 1981, "Fiber-Optic Transducers," *IEEE Spectrum*, pp 24–29.

7. Giallorenzi, T. G., et al., April 1982, "Optical Fiber Sensor Technology," *IEEE Journal of Quantum Electronics*, Vol. Q-E-18, No. 4, pp 626–6.

8. Mellberg, R. S., Summer 1983, "Fiber Optic Sensors," *SRI International*, Research Report No. 684.

Fiber Optic Fundamentals

2

REFRACTION AND TOTAL INTERNAL REFLECTION

Refraction occurs when light passes from one homogeneous isotropic medium to another; the light ray will be bend at the interface between the two media. The mathematic expression that describes the refraction phenomena is known as Snell's Law, which follows:

$$n_0 \sin \phi_0 = n_1 \sin \phi_1 \qquad (2\text{-}1)$$

where

- n_0 = the index of refraction of the medium in which the light is initially travelling
- n_1 = the index of refraction of the second medium
- ϕ_0 = the angle between the incident ray and the normal to the interface
- ϕ_1 = the angle between the refracted ray and the normal to the interface

Figure 2-1 (a) shows the case for light passing from a high index medium to a lower index medium. Even though refraction is occurring, a certain portion of the incident ray is reflected. If the incident ray hits the boundary at ever increasing angles, a value of $\phi_0 = \phi_c$ will be reached, at which no refraction will occur. The angle, ϕ_c, is called the critical angle. The refracted ray of light propagates along the interface, not penetrating into the lower index medium as shown in Figure 2-1 (b). At that point, sin ϕ_c is equal to

Footnote: For $\phi_0 = \phi_c$, $\phi_1 = 90°$; therefore, $\sin \phi_c = n_1 / n_0$.

unity. For angles ϕ_0, greater than ϕ_c, the ray is entirely reflected at the interface and no refraction takes place (see Figure 2-1(c)). This phenomenon is known as total internal reflection (see Figure 2-2).

In Figure 2-2, a ray of light incident upon the end of the fiber at an angle θ will be refracted as it passes into the core. If the ray travels through the high index media at an angle greater than ϕ_c, it will reflect off the cylinder wall, have multiple reflections, and will emerge at the other end of the optical fiber. For a circular fiber, considering only meridional rays (which will be discussed later), the entrance and exit angles are equal. Considering

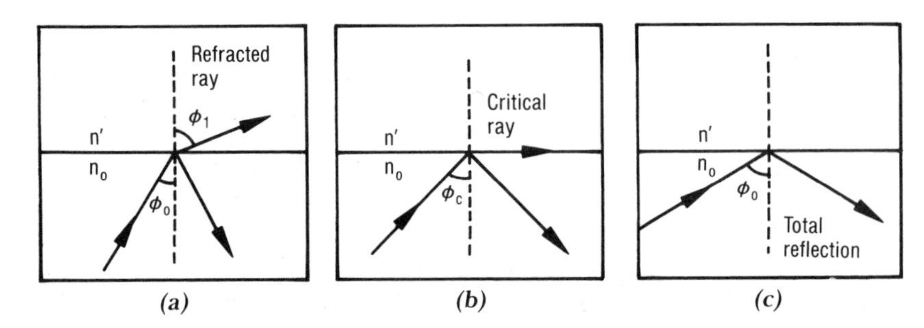

Figure 2-1
Ray Incident at a Plane Interface between a Low and High Index Medium

(reprinted by permission from Academic Press)

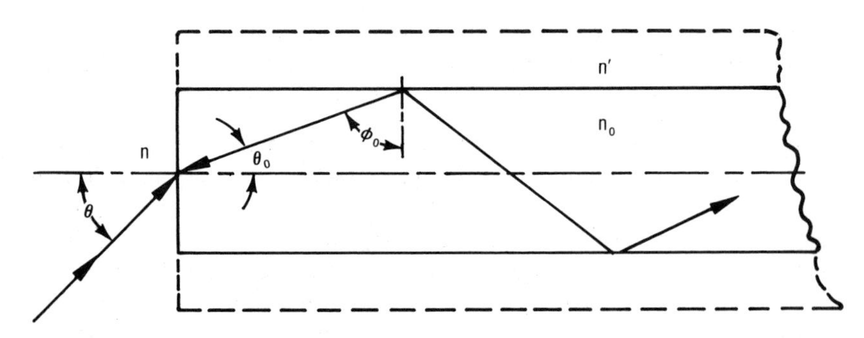

Figure 2-2
Ray Passage along a Flat-Ended Dielectric Cylinder

(reprinted by permission from Academic Press)

Snell's Law for the optical fiber; core index, n_0; cladding index, n_1; and the surrounding media index, n, then:

$$\begin{aligned} n \sin \theta &= n_0 \sin \theta_0 \\ &= n_0 \sin (\pi/2 - \phi_c) \\ &= n_0 [1 - (n_1/n_0)^2]^{1/2} \text{ (see footnote)} \\ &= (n_0^2 - n_1^2)^{1/2} = \text{N.A.} \end{aligned} \quad (2\text{-}2)$$

The quantity $n \sin \theta$ is defined as the numerical aperture (N.A.). The N.A. is determined by the difference in refractive index between the core and the cladding. It is a measure of the light acceptance capability of the optical fiber. As the N.A. increases, so does the ability of the fiber to couple light into the fiber, as shown in Figure 2-3(a). The larger N.A. allows the fiber to couple in light from more severe grazing angles. Coupling efficiency also increases as the fiber diameter increases, since the large fiber can capture more light (see Figure 2-3(b)).[7] Therefore, the maximum light collection efficiency occurs for large diameter core and large N.A. fibers.

MERIDIONAL RAYS

The definitions of total internal reflection and numerical aperture have been based on meridional ray analysis (i.e., the ray path, through its numerous reflections, passes through the longitude axis of the fiber). Using Figure 2-2 as a reference, it can be shown that the length of a meridional ray in a fiber in air ($n = 1$) is:[1]

$$\begin{aligned} l(\theta) &= \frac{L}{[1 - (\sin \theta / n_0)^2]^{1/2}} \\ &= L \sec \theta_0 \end{aligned} \quad (2\text{-}3)$$

where $l(\theta)$ is the length of the optical path for a ray inclined to the fiber axis at an angle θ_0, and L is the length of the fiber measured along the fiber axis. It is interesting to note that the length of the optical path for an incident ray depends only upon the fiber length, the angle of incidence, and the refractive index of fiber core, while being independent of fiber diameter. The number of reflections, however, is dependent upon fiber diameter. The larger the diameter of the fiber, the fewer the number of reflections as defined below:[1]

$$N = (L/d) \tan \theta_0 \quad (2\text{-}4)$$

where N is the number of reflections at the core/cladding interface and d is the fiber diameter.

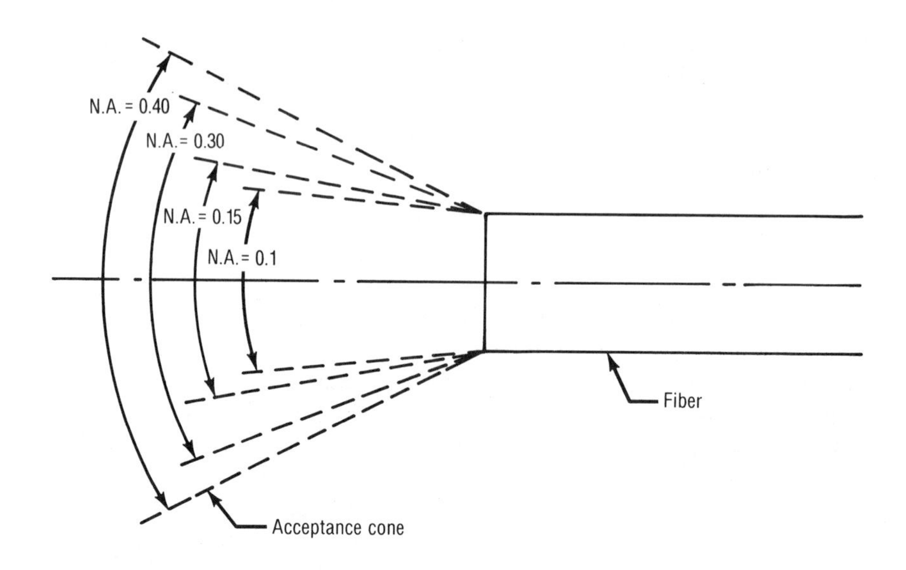

(a) Numerical aperture

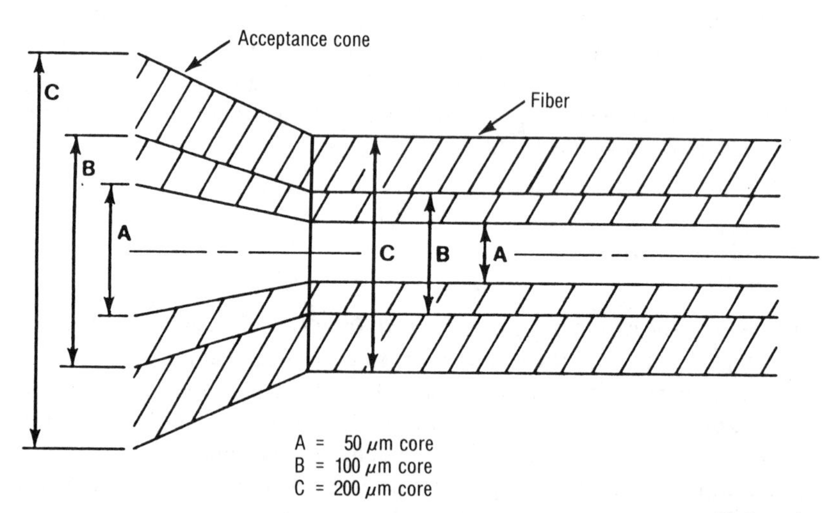

(b) Core size

Figure 2-3
Acceptance Angle Associated with Numerical Aperture and Core Size

(reprinted by permission from Instrument Society of America)

SKEW RAYS

In general, meridional rays describe a very simplified ray propagation. Often rays are skewed in nature and dominate the optical properties of the fiber. These rays affect the "real" ray propagation and alter the simplified definition of numerical aperture, ray path length, and number of reflections.

Skew rays travel along a helical path as illustrated in Figure 2-4. For simplicity, however, this book will use meridional ray analysis. A more detailed treatment of skew rays is given in Reference 1.

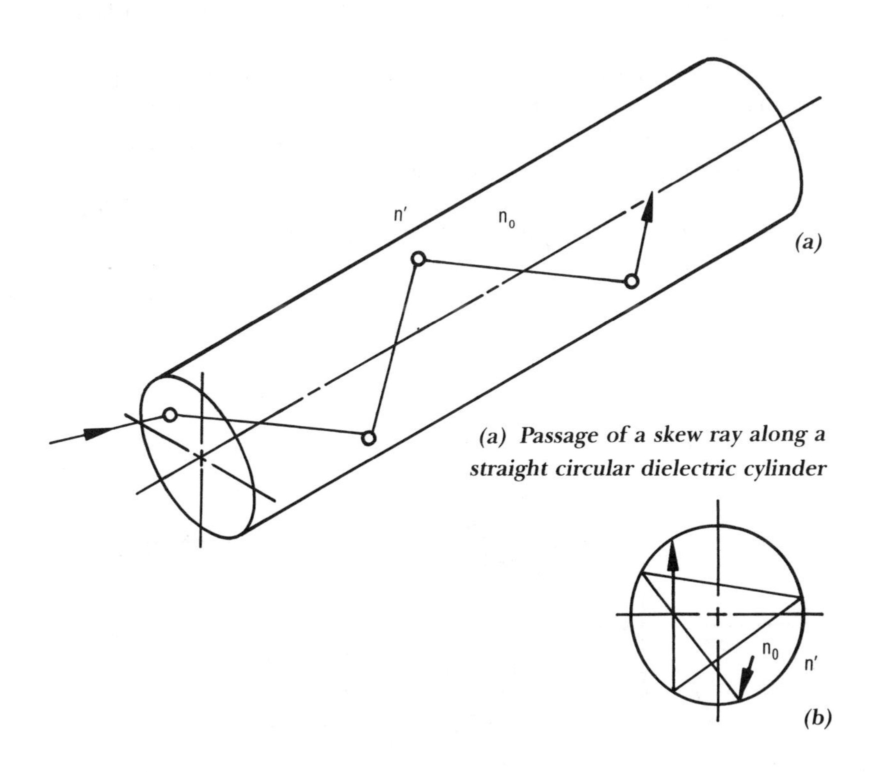

(a) Passage of a skew ray along a straight circular dielectric cylinder

(b) Helical path traversed by the skew ray

Figure 2-4
Skew Rays

(reprinted by permission from Academic Press)

BENT FIBERS

Figure 2-5 shows clearly the effect of bending on a meridional ray travelling in a fiber. For a straight fiber, the angle between the light ray and the normal to the plane of reflection is defined by the angle ϕ. However, as shown in Figure 2-5, when the fiber is bent, the plane of reflection and the reflective ray rotate by the angle δ. Therefore, for a curved fiber, the angle between the reflected ray and the tangent at the reflection point is $\phi - \delta$. In a straight fiber, for $\phi > \phi_c$, the rays will be totally internally reflected. In a bent fiber, the effective critical angle is reduced by δ. Therefore, rays incident between ϕ_c and $\phi_c - \delta$ will be lost through the cladding of the fiber. The effective critical angle is reduced in a bent fiber, and the amount of light that can be injected into the fiber is reduced.

Figure 2-6 illustrates a curved fiber with rays traced through several internal reflections. The multi-reflections in a bent fiber with a constant radius of curvature do not increase the bending loss since the steepness of the rays as they hit the side walls of the fiber does not increase (i.e., $\phi_1 = \phi_2 = \phi_3$). The bending loss occurs primarily at the transition from the straight to the bent section (a change in one curvature to another curvature for multiple bending points).

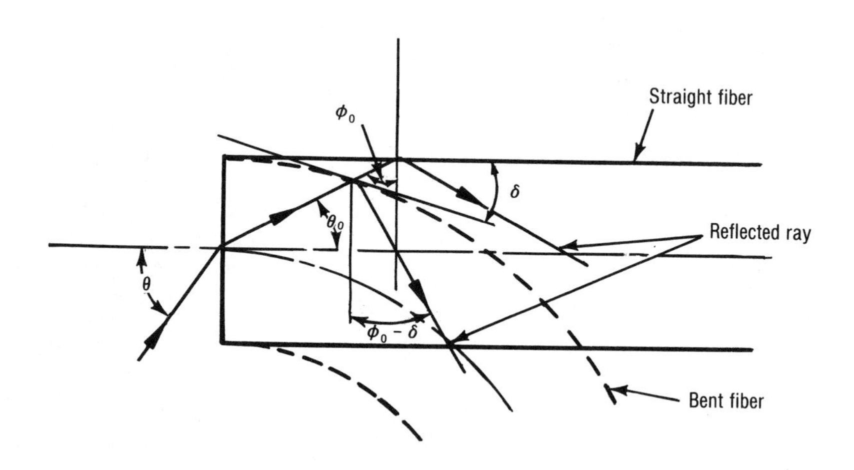

Figure 2-5
Bending Effects on Critical Angle

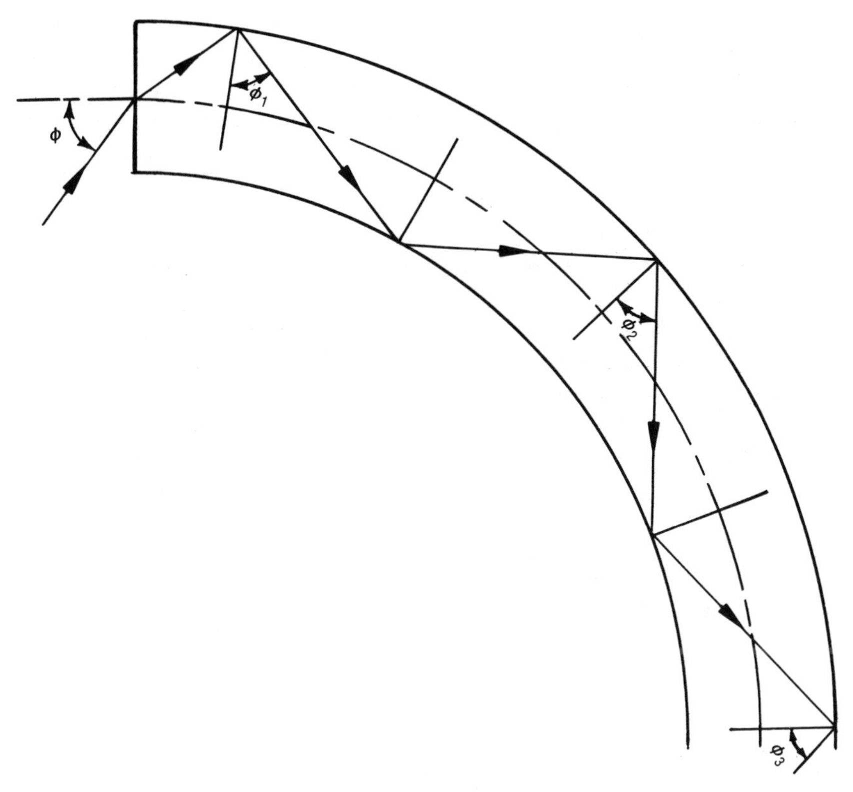

Figure 2-6
Multiple Reflections in a Bent Fiber

MECHANISMS OF ATTENUATION

Attenuation or loss is defined by the following equation:

$$A = -10 \log P_i / P_o \tag{2-5}$$

where A is the attenuation and P_i and P_o are the input power and output power, respectively. The negative sign arises from the convention that attenuation is negative. Attenuation is measured in decibels (dB) per unit length, typically dB/km.

Loss can vary from 1 to 1000 dB/km in useful fibers with the various causes of loss often being wavelength dependent. The causes for loss are absorption, scattering, microbending, and end loss due to reflection, as shown in Figure 2-7.[2]

Typically, fibers have contaminations of transition metal such as copper, chromium, iron, and hydroxyl ions. Figure 2-8 shows a typical fiber loss curve for a telecommunication grade fiber and the effect of various attenuation mechanisms. The cumulative effects of iron and copper contamination

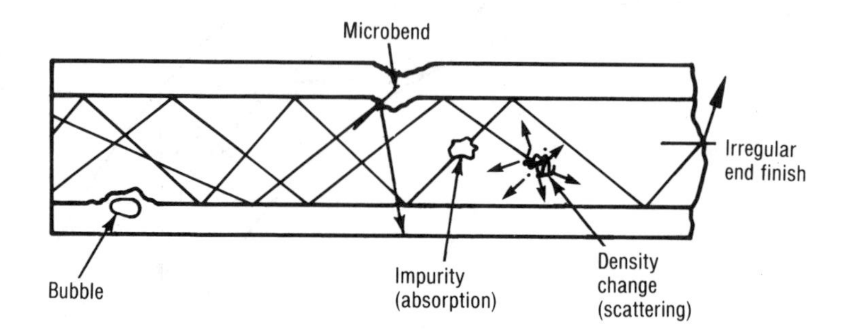

Figure 2-7
Causes for Attenuation

(reprinted by permission from AMP Inc.)

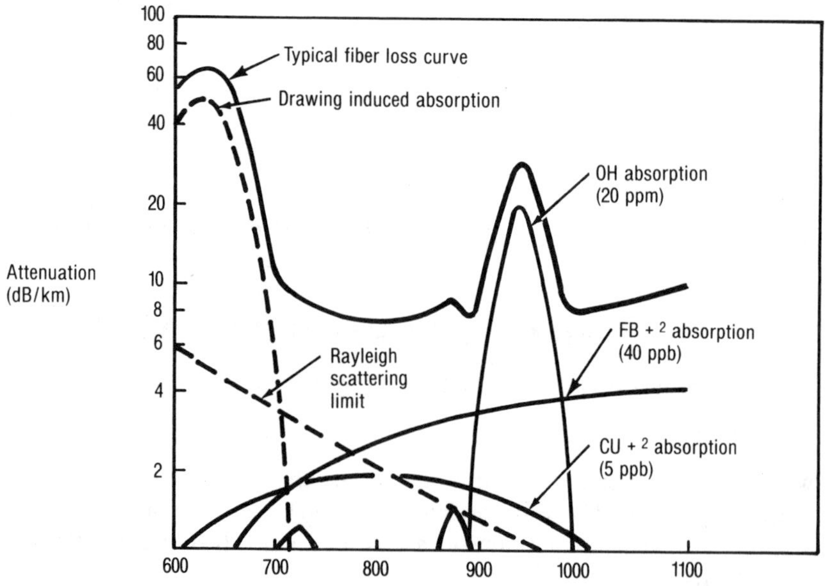

Figure 2-8
Absorption and Scattering are Functions of Wavelength

(reprinted by permission from AMP Inc.)

and scattering result in a curve having wavelength "windows" in which transmission is optimized. Until recently, most telecommunication fibers were used at 820 nm (see Figure 2-8).

As shown in Figure 2-9, the attenuation can be further reduced at high wavelengths if a low $(OH)^-$ content is achieved. Fibers for telecommunication applications are moving to higher wavelength optimized "windows" in the 1100 nm to 1500 nm range. The loss decreases at higher wavelengths because inherent Rayleigh scattering losses are decreased, since the scattering is proportioned to $1/\lambda^4$ where λ is the wavelength. Also, a second transmission window occurs if the $(OH)^-$ content is reduced. Generally, it is not possible to completely remove the $(OH)^-$ absorption, and an absorption peak remains. Therefore, attenuation curves as shown in Figure 2-9 are referred to as double "window" fibers (i.e., two windows for low loss operation).

Sensors are often required to work over a broad wavelength range. Figure 2-10(a)[3] shows a transmission versus wavelength curve for synthetic silica. The transmission range is 180 nm in the UV to over 2 microns in the near IR. If the water content is reduced [less $(OH)^-$], the infrared transmission is improved (see Figure 2-10(b)). Bundle fibers, which are very useful in

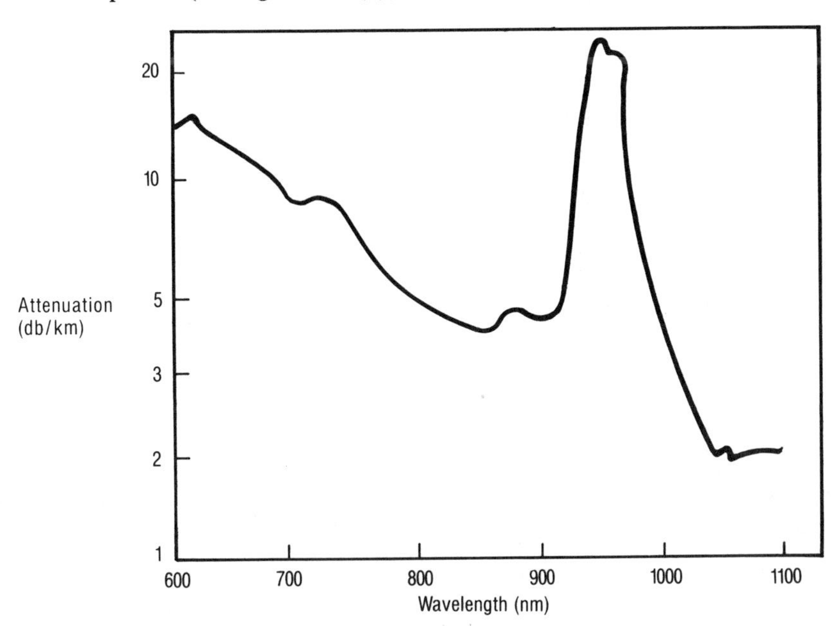

Figure 2-9
Attenuation Curve for a Typical Low-Loss Fiber

(reprinted by permission from AMP Inc.)

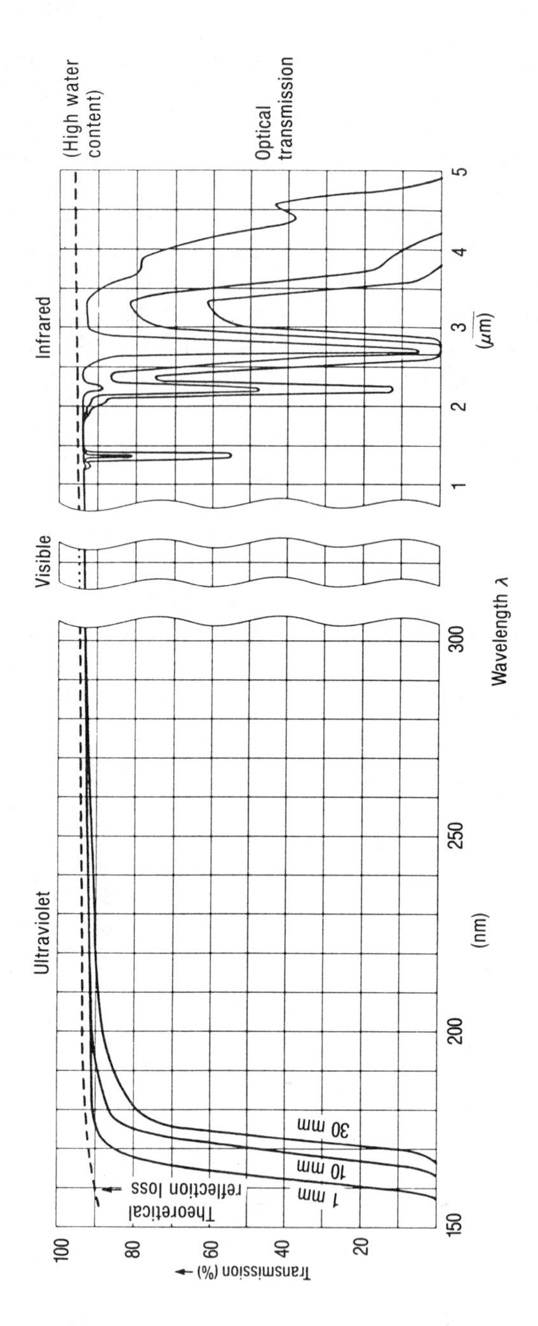

Figure 2-10(a)
Transmission Characteristics of Silica
(reprinted by permission from Heraeus AMERSIL Inc.)

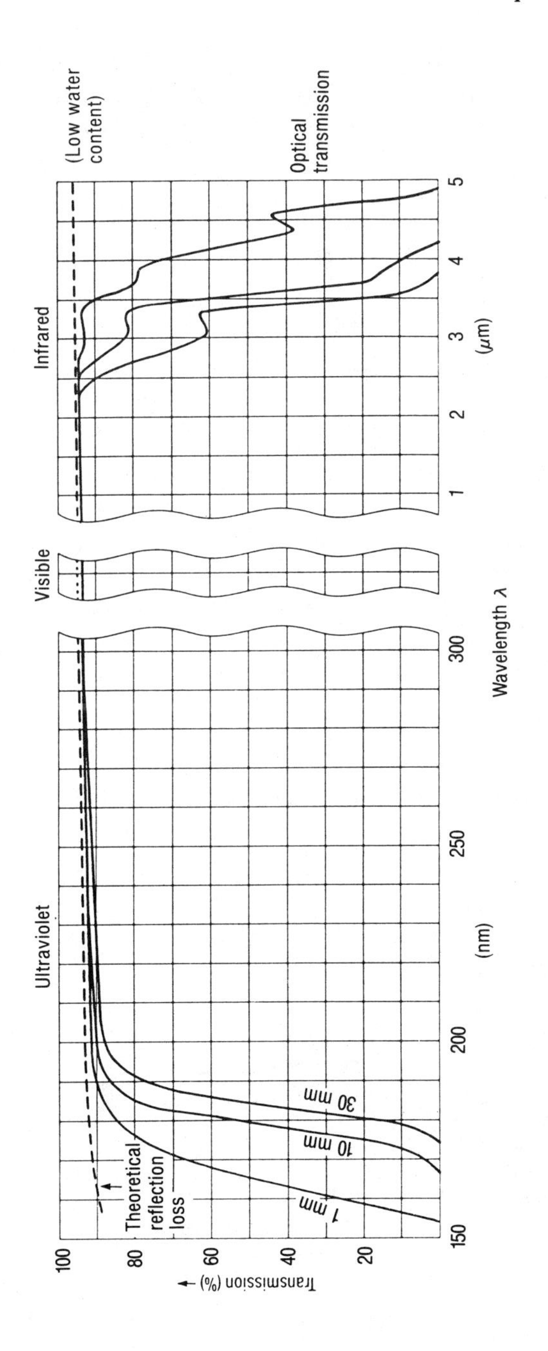

Figure 2-10(b)
Transmission Characteristics of Silica
(reprinted by permission from Heraeus AMERSIL Inc.)

sensors, have typical transmission vs. wavelength curves as shown in Figure 2-11.[8] The transmission level is approximately 400–600 dB/km. In most applications, bundle fibers are limited to 10 to 25 feet before loss becomes excessive. The loss in bundle fiber is associated with high transition metal concentrations in the fibers.

Losses associated with microbending in the fiber will be discussed in depth in a later chapter since the microbending mechanism is quite useful in sensor design.

The following definition of fiber loss is useful.[2]

Low loss — Less than 10 dB/km
Medium loss — 10 to 100 dB/km
High loss — Greater than 100 dB/km

In addition to losses in the fiber itself, there are losses at the ends of the fiber due to reflection. The refractive index difference between the fiber and, usually, an interface leads to Fresnel reflection losses. As a result, small amounts of energy are reflected back into the fiber. These losses show up in connecting the fiber to optical devices or other fibers and must be considered in overall system losses.

The Fresnel reflection loss, R, is defined for a glass air interface by:

$$R = \left[\frac{n_0 - 1}{n_0 + 1}\right]^2 \qquad (2\text{-}6)$$

where n_0 is the index of refraction of the core material. Using the classical definition of absorption:

$$P_o = P_i e^{-\alpha L} \qquad (2\text{-}7)$$

where P_o and P_i are the output power and input power, respectively, α is the

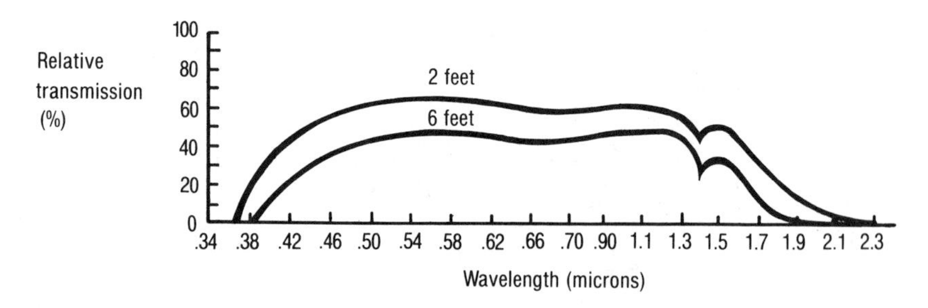

Figure 2-11
Transmission versus Wavelength for Bundle Fiber
(reprinted by permission from Welch Allyn Inc.)

attenuation coefficient (units/length), and L is the length. Transmission, T, in percent, is given by:

$$T = (1 - R)^2 e^{-\alpha L} \tag{2-8}$$

The term $(1 - R)^2$ is the reflection loss for the entrance and exit ends of the fiber. The effect of reflections is multiplicative and therefore accounts for the square term since there are two surfaces (exit and entrance).

WAVE GUIDE PROPAGATION

For the case of sensors, most fibers have a step index profile (shown in Figure 2-16). The step fiber has an index of refraction constant in the core, constant in the cladding, and discontinuous at the interface. Depending upon the core size and numerical aperture, the fiber will transmit many modes (rays) of light and be referred to as multimode fiber. It may also be limited to a single mode.

Modal performance is mathematically defined by Maxwell's Equation for cylindrical boundary conditions as follows.[4,5]

$$\frac{d^2\psi}{d\rho^2} + \frac{1}{\rho}\frac{d\psi}{d\rho} + \frac{1}{\rho^2}\frac{d\psi}{d\phi} + (k^2 - \beta^2) = 0 \tag{2-9}$$

where

ρ = the radial parameter
ψ = the wave function of the guided light
k = the bulk medium wave vector
β = the wave vector along the fiber axis
ϕ = the azimuthal angle

If the wave function is assumed to be of the form:

$$\psi = AF(\rho)\, e^{i\nu\phi} \tag{2-10}$$

then the Maxwell Equation becomes a Bessel Equation. The boundary conditions require that on the axis ($\rho = 0$) the field has a finite value. However, the field becomes zero at infinity ($\rho = \infty$). The resulting longitudinal component of the field functions are as follows:

$$AJ_\nu\,(ur/a)\, e^{i\nu\phi},\ \rho < a \text{ (core)}$$

$$BK_\nu\,(wr/a)\, e^{i\nu\phi},\ \rho > a \text{ (cladding)} \tag{2-11}$$

where $J_\nu\,(ur/a)$ and $K_\nu\,(wr/a)$ are Bessel functions of the first and second kind, respectively:

$U^2 = (k_1^2 - \beta^2)\, a^2$

$k_1 = 2\, \pi\, n_1 / \lambda_0$

$w^2 = (\beta^2 - k^2_2)\, a^2$

$k_2 = 2\, \pi\, n_2 / \lambda_0$

The subscripts 1 and 2 denote the core and the cladding, respectively, while a is the radius of the core.

$$w^2 + u^2 = (2\, \pi\, a/\lambda_0)^2\, (n_1^2 - n_2^2) = V^2 \tag{2-12}$$

The term $w^2 + u^2$ is a constant for all modes and is characteristic of the optical fiber. The parameter V represents the number of modes in the fiber and is related to the numerical aperture as follows (see Figure 2-12):

$$V^2 = \frac{2\, \pi\, a}{\lambda_0}\, (\text{N.A.})^2 \tag{2-13}$$

The relationship clearly follows the previously developed concept for numerical aperture (i.e., as the N.A. increases, so does the number of rays (modes) that can be accepted). It is important to note that the mathematical solutions to Maxwell's Equation have allowed values; therefore, there are allowed modes. Modes that do not fit the mathematical solutions are not

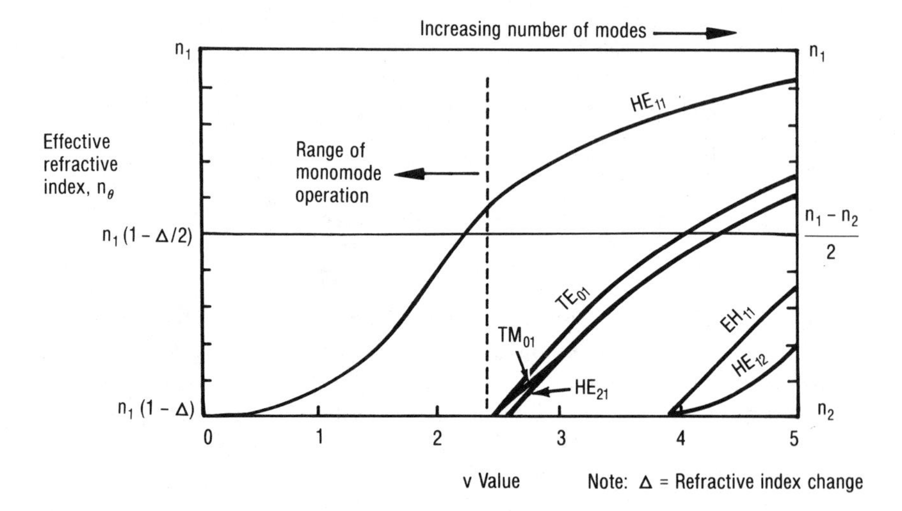

Figure 2-12
Mode Formation with V Value

(reprinted by permission from Garland STPM Press, © 1979)

allowed. As a result, modes can be considered as being quantized. For the simplest case of a single-mode fiber $\nu = 0$, only the *TE* and *TM* modes are present. Two modes exist in a single-mode fiber because the mode can degenerate due to polarization. For higher modes where $\nu \neq 0$, the modes contain x, y, z field components, which are referred to by the notation HE_{mn} $(n = \nu)$ or EH_{mn}, depending upon the dominance of magnetic or electric characteristics. The subscript n defines further mathematical solutions due to the behavior of Bessel functions. The field varies in a periodic fashion with ϕ and ρ. Skew rays that have a ϕ component result in a power concentration away from the fiber axis near the cladding.

At $V < 2.405$, the fiber can support only a single mode, designated as HE_{11}. At $V > 2.405$, other modes can exist, with the number increasing as V increases. Each of the modes is doubly degenerate due to polarization since, in circular waveguides, two orthogonal polarization states (modes) exist for the same wave number. As will be discussed later, elliptical core geometric considerations can eliminate the degeneracy.

It is clear from Equation 2-13 that single-mode transmission requires small core size and low values of numerical aperture. However, if the V value is small due to a very low N.A., the single ray is loosely guided and susceptible to bending losses.

Figure 2-13 is a plot of wavelength vs. attenuation for single-mode and multimode fibers.[5] Single-mode fibers have a strong wavelength dependence, while multimode fibers are relatively insensitive to wavelength changes. For a single-mode fiber, as λ_0 increases, V decreases; the ray is less tightly guided and tends to enter the cladding where it is scattered. As λ_0 decreases, V increases; the light shifts to higher order modes and is lost. Multimode fibers in which hundreds of modes are propagating can tolerate modal shifts over a wide wavelength range (see Figure 2-12).

EVANESCENT WAVE

Energy is carried in the cladding as well as the core of the fiber. This is referred to as the evanescent wave phenomena. There is an exponential loss of the energy in the fiber cladding as the distance from the fiber core is increased. Figure 2-14(a) shows a wave incident upon the core cladding interface. The wave fronts (planes of constant phase) are shown as dotted lines. The arrows represent the wave normals. The direction of the arrows indicates the flow of energy. The superposition of incident and reflected beams results in a wave where the wavefront is perpendicular to the interface, as shown in Figure 2-14(b). The energy distribution in the core has its

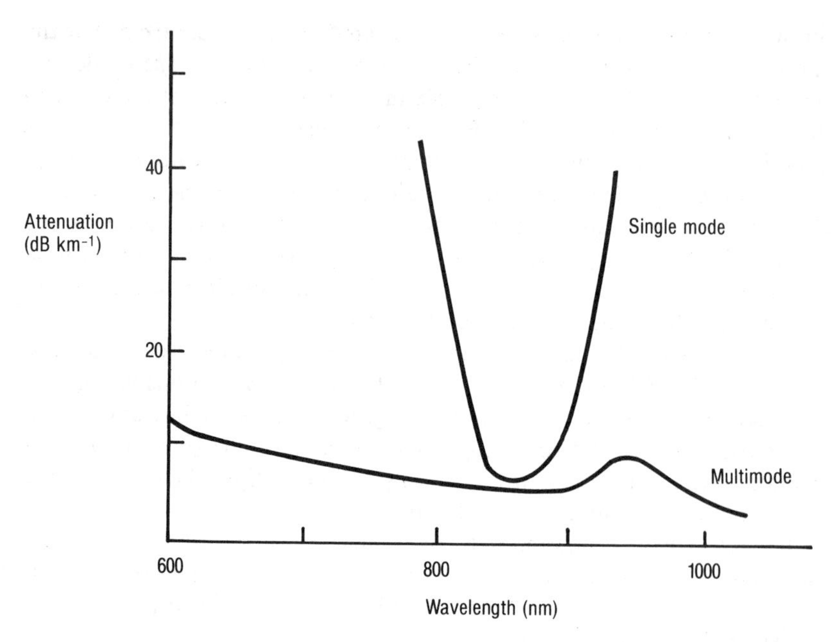

Figure 2-13
Fiber Attenuation as a Function of Wavelength
(© 1978 IEEE; reprinted with permission)

maximum near the interface. The energy in the cladding falls off exponentially.

CROSS COUPLING

Cross coupling involves the transfer of light from one adjacent fiber to another. The extent of coupling between two adjacent fibers in close proximity is a function of the following: the diameter of the fiber; the spacing between fibers; the length in which the fibers are in close proximity; the index of the fiber core, the cladding and the index of surrounding media; the thickness of the cladding; the wavelength of light in the fiber; and the particular modes excited in the fiber. Clearly, a precise mathematic expression for the phenomenon is quite complex.(1)

It is possible by properly selecting the various parameters to completely transfer all the energy in one fiber to an adjacent fiber. The length required

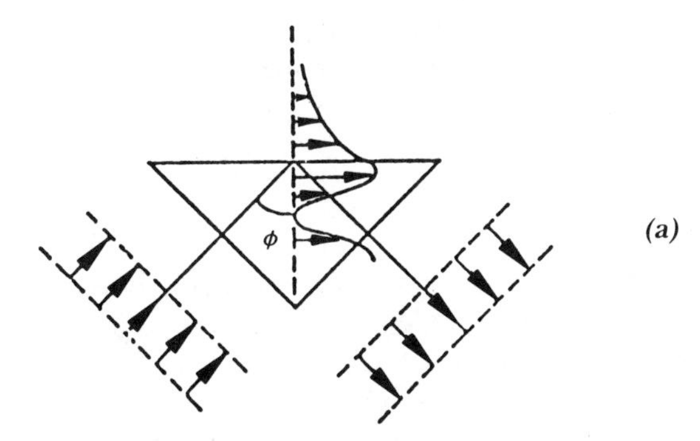

(a) The evanescent wave phenomenon at plane interface

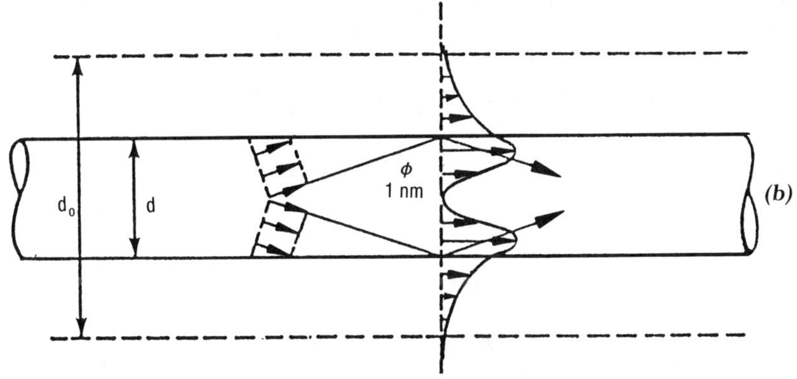

(b) Evanescent wave and interference phenomena at cylindrical fiber boundary

Figure 2-14
Evanescent Wave Phenomenon
(reprinted by permission from Academic Press)

for the transfer is referred to as the "beatlength".* For large fiber diameters, the energy transfer is limited over some length; a state of equilibrium will be reached. For small fibers, when the radiation is coherent, almost all the energy will be transferred from one fiber to the other in a length $X/2$ and then return to the original fiber in an $X/2$ interval, where X is the beatlength. This phenomenon is known as beating.

---

*Beatlength is also used to describe observed light and dark bands in polarizing maintaining fiber, which is discussed in a later chapter.

It has been found that as the wavelength increases, the beatlength decreases. As a result, the relative intensity of energy in adjacent coupling fibers will vary with wavelength, as shown in Figure 2-15.

MODE PATTERNS

As discussed previously, not all rays are allowed to propagate in an optical fiber because certain boundary conditions and Maxwell's Equation must be satisfied. Therefore, if the light emitting from a fiber were projected on a screen, it would not give a uniform illumination even for a fiber with only one or a few modes. Rather, a geometrical pattern of light and dark is observed, corresponding to the complex energy distribution associated with the various modes. For fibers with large N.A. values, the individual modes are not visible due to modal interaction, but distinct patterns are possible such as concentric rings and the so-called "speckle pattern". The use of the

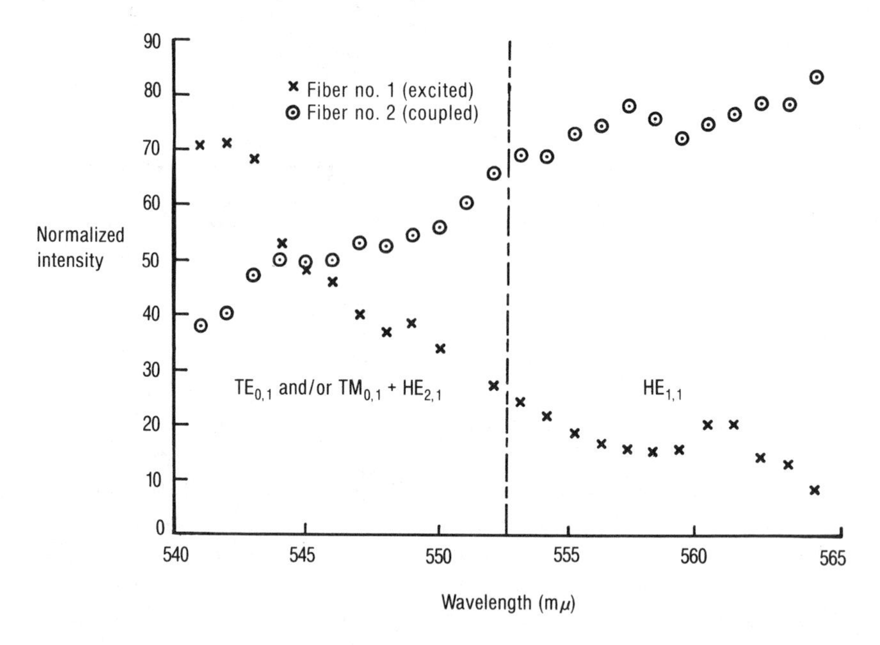

Figure 2-15
Plot Intensity of the Excited Fiber and the Coupled Fiber as a Function of Wavelength of Incident Radiation

(reprinted by permission from Academic Press)

mode patterns has had limited use in sensor technology. However, a more detailed analysis is given by Kapany.[1]

The amount of energy carried in the cladding depends upon the excited modes in the core. As the modes that are closer to the critical angle are excited, more energy is propagated in the cladding.

FIBER TYPES

Figure 2-16 shows characteristics of the three basic fiber configurations.[2] The multimode step index and single-mode step index fibers have similar refractive index profiles, but the single-mode fiber has a much smaller core. The single-mode fiber has much less dispersion and a sharp output pulse. The graded index fiber has an intermediate output pulse.

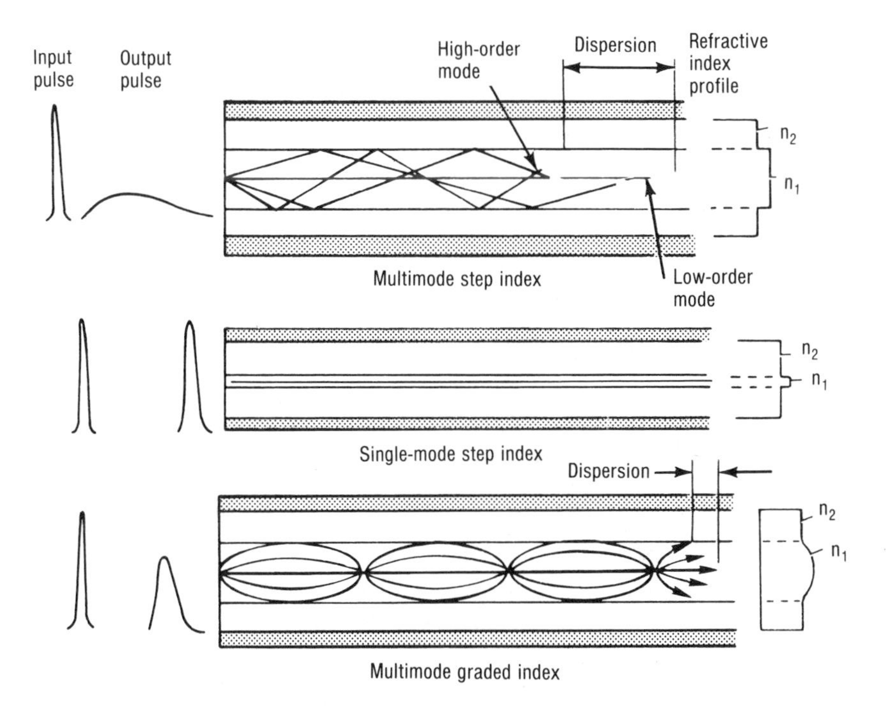

Figure 2-16
Core Size and Refractive Index Determine the Light Propagation Characteristics of the Fiber

'(reprinted by permission from AMP Inc.)

The multimode graded index fibers are used to increase bandwidth (i.e., information carrying capacity) as compared to multimode step index fibers. However, they will not play a great role in sensor technology. Most of the emphasis will be on step index fibers, both single-mode and multimode. Step index fibers have an index of refraction that is constant in the core and the cladding and stepped at the core/cladding interface. Table 2-1 lists some of the fiber types and their various properties.

References 9-15 will also be useful as background information.

Table 2-1
Typical Properties of Various Fiber Types

Type	Loss (dB/km)	N.A.	Core Size (O.D. Microns)	Core-to-Cladding Ratio	Bandwidth (MHz-km)
Multimode Step Index					
Glass-Clad Glass (Bundle)	400–600	0.4–0.6	(50–70)	0.9–0.95	20
Plastic-Clad Silica	3–10	0.3–0.4	200–600	0.7	20
Glass-Clad Glass (Low Loss)	2–6	0.2–0.3	50–200	0.4–0.8	20
Single Mode	2–6	0.15	5–8	0.04	1000

REFERENCES

1. Kapany, M. S., 1967, *Fiber Optics — Principles and Applications*, Academic Press, New York.
2. Anonymous, 1982, *Designers Guide to Fiber Optics*, AMP, Inc., Harrisburg, PA.
3. Product Specification — Heraeus Amersil.
4. Davis, C. M., et al., 1982, *Fiber Optic Sensor Technology Handbook*, Dynamic Systems, Reston, Virginia.
5. Giallorenzi, T. G., July 1978, "Optical Communications Research and Technology: Fiber Optics," *Proceedings of the IEEE*, Vol. 66, No. 7, pp 744–80.
6. Boyle, W. S., August 1977, "Light-Wave Communications," *Scientific American*, Vol. 237, No. 2, pp 40–48.

7. Coulombe, R. F., 1985, "A Practical Approach to Plant Data Bus Networks Utilizing Fiber Optics," *Proceedings of the ISA International Conference*, Philadelphia, PA, Vol. 40, Part 2, pp 1027–33.
8. Product Specification — Welch Allyn.
9. Wolf, H. F., 1979, *Handbook of Fiber Optics — Theory and Applications*, Garland STPM Press, New York, p. 63.
10. Snyder, A. W., and Love, J. D., 1983, *Optical Waveguide Theory*, Chapman and Hall, New York.
11. Midwinter, J. E., 1970, *Optical Fibers for Transmission*, Wiley, New York.
12. Marcuse, Dietrich, 1981, *Principles of Optical Fiber Measurements*, Academic Press, New York.
13. Poly, James C. (Ed.), 1984, *Fiber Optics*, CRC Press, Boca Raton, FL.
14. Gowar, John, 1984, *Optical Communications Systems*, Prentice/Hall International, Englewood Cliffs, NJ.
15. Senior, John M., 1985, *Optical Fiber Communications: Principles and Practice*, Hall International, Englewood Cliffs, NJ.

Intensity-Modulated Sensors

3

INTRODUCTION

Intensity-modulated sensors were defined in Chapter 1 as sensors that detect the variation of the intensity of light associated with the perturbing environment. The general concepts associated with intensity modulation include transmission, reflection, and microbending. However, several other mechanisms that can be used independently (intrinsically) or in conjunction with the three primary concepts include absorption, scattering, fluorescence, polarization, and optical gratings. While intensity-modulated sensors are analog in nature, they have significant usage in digital (on/off) applications for switches and counters.

TRANSMISSIVE CONCEPT

The transmissive sensor concept[1,2] is normally associated with the interruption of a light beam in a switch configuration. However, this approach can provide a good analog sensor. Figure 3-1 (a) shows the probe configuration for measurement of axial displacement. Figure 3-1 (c) gives a curve of output versus distance between the probes. The curve follows a $1/r^2$ law where r is distance. A more sensitive transmissive approach employs radial displacement as shown in Figure 3-1 (b). The sensor shows no transmission if the probes are displaced a distance equal to one probe diameter. Approximately the first 20% of the displacement gives a linear output. The curve in

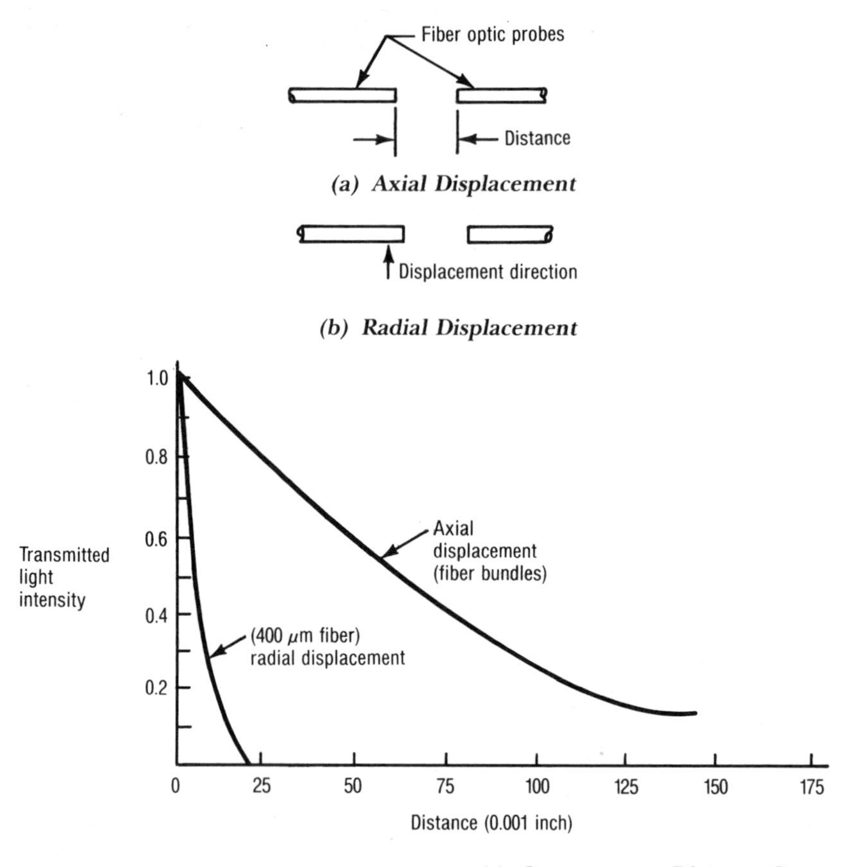

Figure 3-1
Transmissive Fiber Optic Sensor

Figure 3-1, showing the effects of radial displacement, is for probes with single fibers, 400 microns in diameter.

A modification of the transmissive concept is referred to as frustrated total internal reflection.(3) The two opposing probes have the fibers polished at an angle to the fiber axis, which produces total internal reflection for all propagating modes, as shown in Figure 3-2. As the fiber ends come close in proximity to one another, energy is coupled. The intensity of light coupled into the receiving fiber is shown in Figure 3-3. This approach provides the highest sensitivity for a transmissive sensor.

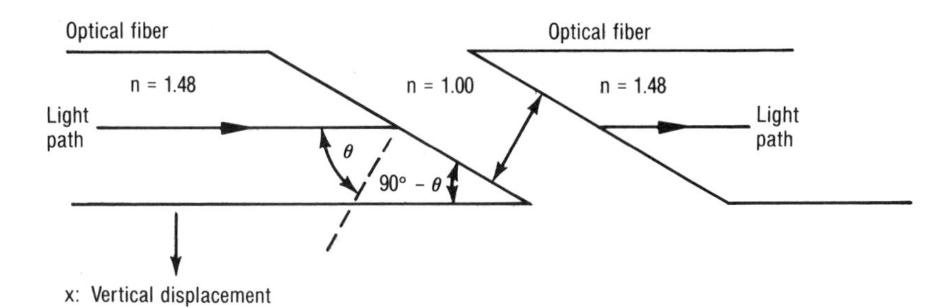

Figure 3-2
Frustrated Total Internal Reflection Configuration
(reprinted by permission from Optical Society of America)

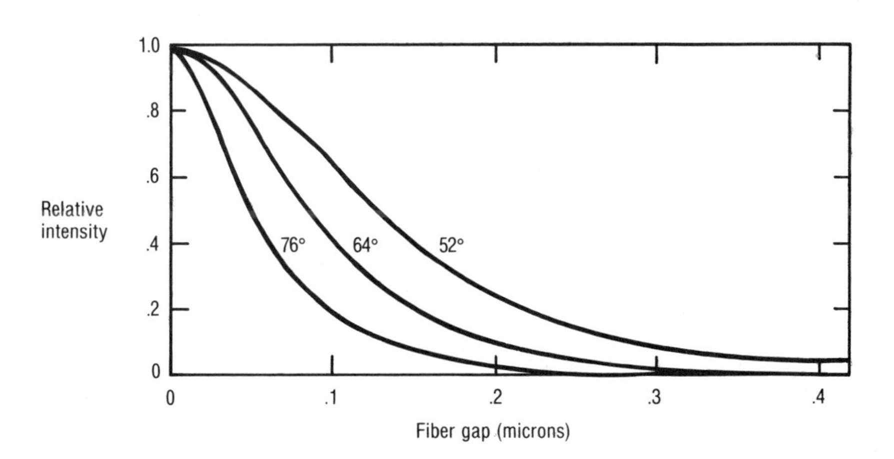

Figure 3-3
Frustrated Total Internal Reflection Response Curves
(reprinted by permission from Optical Society of America)

REFLECTIVE CONCEPT

The reflective concept(4,5) is especially attractive for broad sensor use due to accuracy, simplicity, and potential low cost. The concept is shown in Figure 3-4(a). The sensor is comprised of two bundles of fibers or a pair of single fibers. One bundle of fibers transmits light to a reflecting target, the other bundle traps reflected light and transmits it to a detector. The intensity of the detected light depends on how far the reflecting target is from the fiber optic probe. Figure 3-4(b) shows the detected light intensity versus distance from the target. The linear front slope allows a displacement to be measured with potential accuracy of one millionth of an inch. The accuracy depends on the probe configuration; a hemispherical probe has more dynamic range, but less sensitivity when compared to a random probe (Figure 3-4). A fiber pair probe further expands the dynamic range. A single fiber used in conjunction with a beam splitter to separate the transmitted and the received beams eliminates the front slope. Depending upon the fiber configuration, reflective probes can be tailored for a wide range of applications.

For applications that require a greater dynamic range than possible with any of the fiber configurations, a lens system can be added[6] as shown in Figure 3-5(a). Using a lens system in conjunction with a fiber optic probe, the dynamic range can be expanded from 0.2 inch to 5 or more inches as shown in Figure 3-5(b). The detailed mechanism of how the lens extends the sensor range is described in Chapter 6 on displacement measurement.

MICROBENDING CONCEPT

Another attractive sensor concept is that of microbending.[7,8,9] If a fiber is bent, small amounts of light are lost through the wall of the fiber. If a transducer bends the fiber due to a change in some physical property, as shown in Figure 3-6(a), then the amount of received light is related to the value of this physical property. Figure 3-6(b) indicates that as pressure causes the transducer to squeeze together and bend the fiber, the amount of transmitted light decreases with displacement. Like reflective sensors, they are potentially low cost and accurate. It is also important to note that microbending sensors have a closed optical path and, therefore, are immune to dirty environments.

In the response curve, the nonlinear behavior initially is due to, at least in part, the rheological behavior of the polymeric protective coatings. The change in slope at high displacement values is due to light depletion. The linear central portion of the curve is the active sensing region. In general, as the number of bend points increase and/or the spacing between bend points decreases, the sensitivity increases.

INTRINSIC CONCEPT

Intrinsic sensors change the intensity of the returning light from the sensor, but, unlike the transmissive, reflective, and microbending concepts, no movement is required. Intrinsic sensors use the chemistry of the core glass

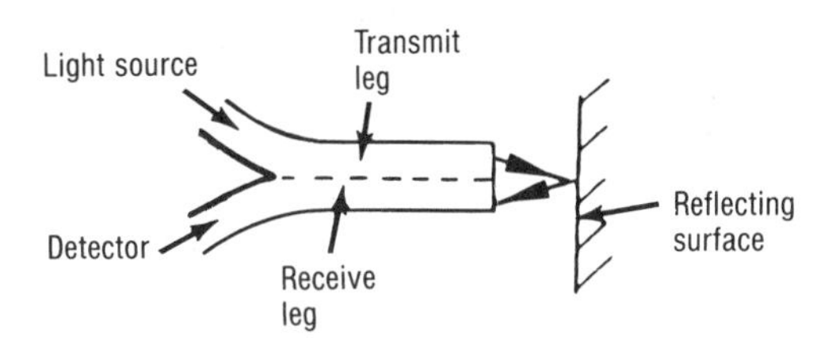

(a) Sensor Arrangement

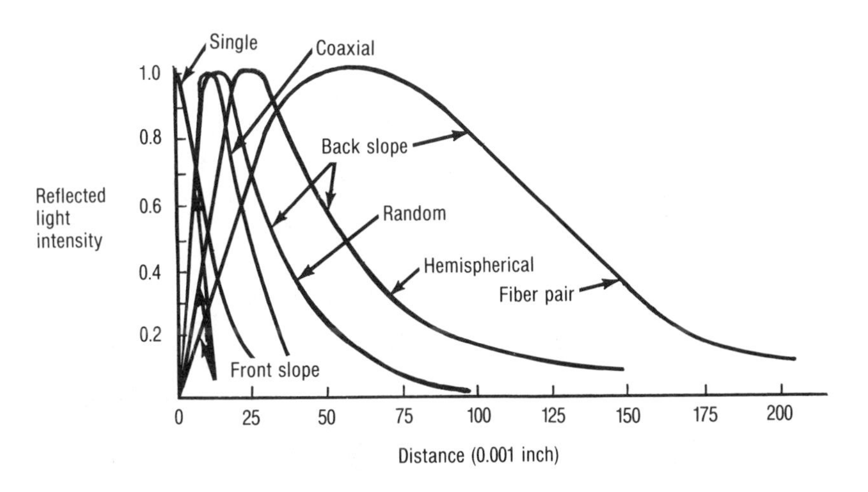

(b) Output versus Distance

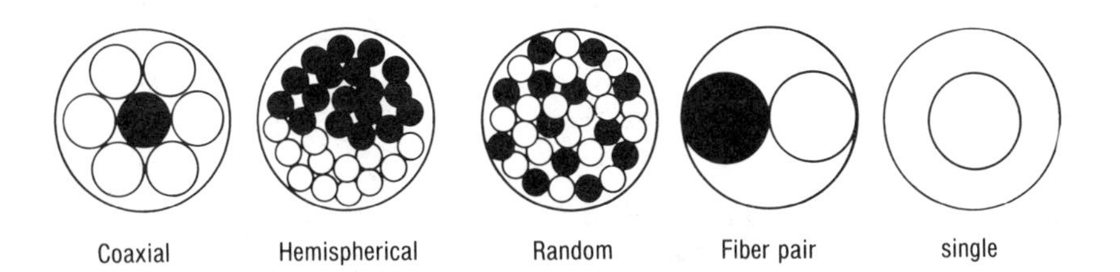

(c) Probe Configuration

Figure 3-4
Reflective Fiber Optic Sensor Response Curve for Various Configurations

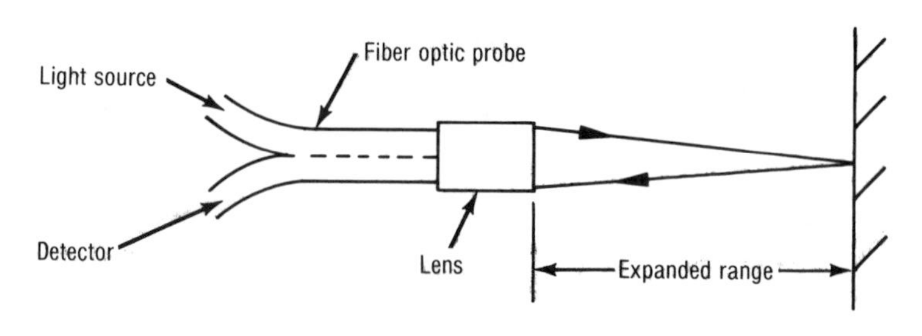

(a) Configuration

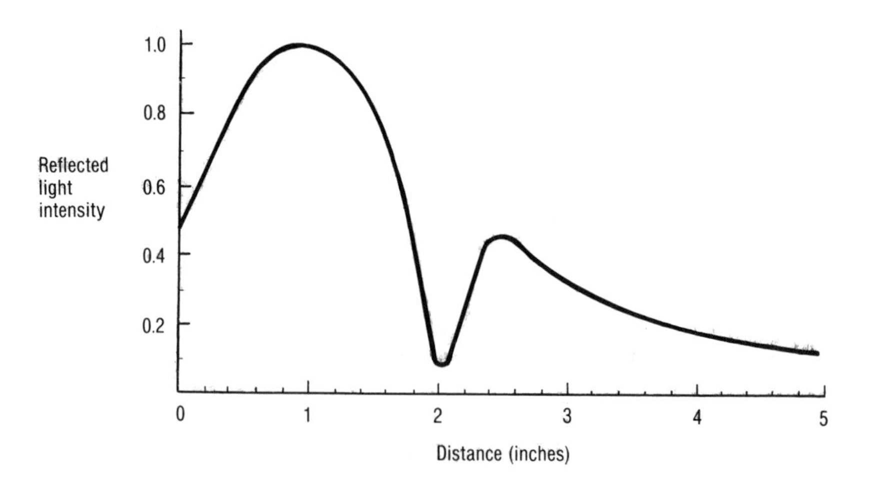

(b) Response Curve

Figure 3-5
Reflective Fiber Optic Sensor with Expanded Range

(cladding glass or the plastic coatings) to achieve the sensing activity. The prime mechanisms are absorption, scattering, fluorescence, changes in refractive index, or polarization.

For absorption, doping the core glass results in absorption spectra.(10) Generally, some peaks are temperature sensitive, while others are not. The ratio of intensity at two specified wavelengths provides a temperature sensing function as shown in Figure 3-7. A similar approach can be considered for scattering.

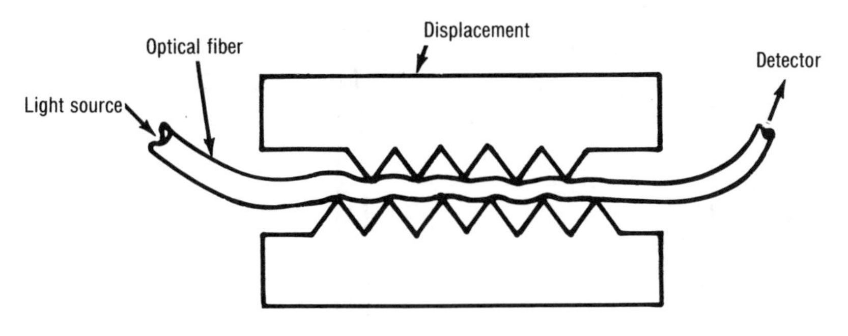

(a) Sensor Arrangement

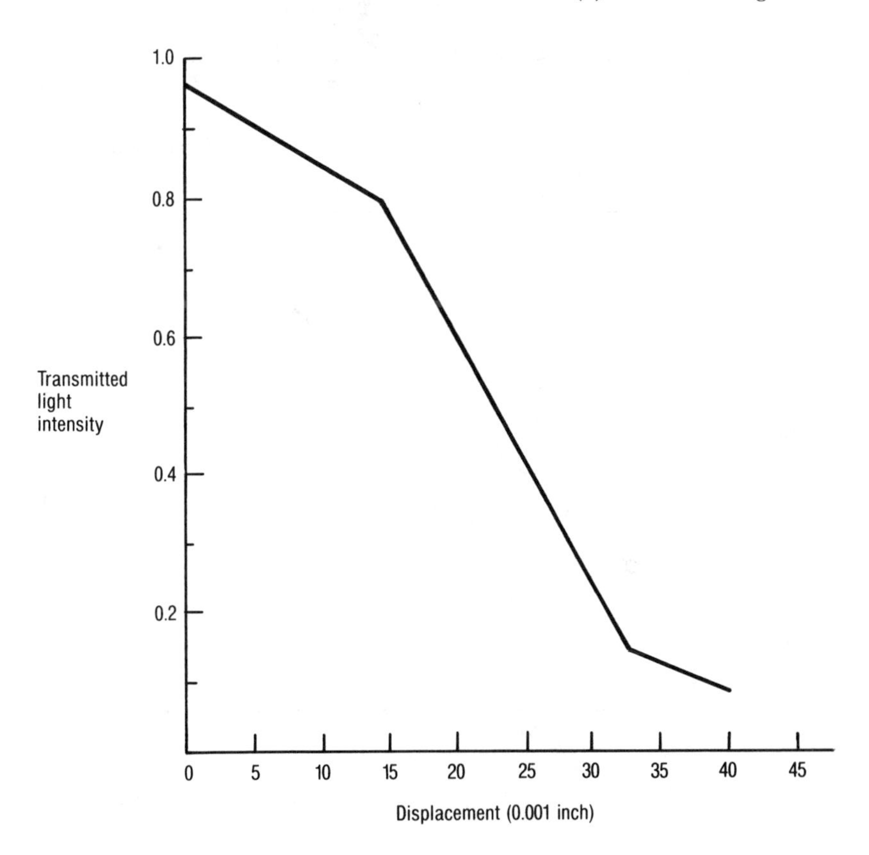

(b) Output versus Displacement

Figure 3-6
Microbending Sensor

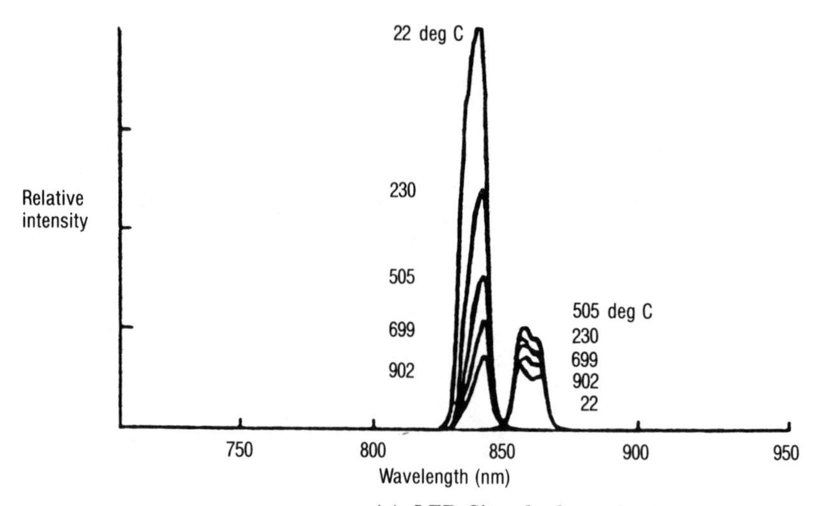

(a) LED Signals through Nd Doped Fiber

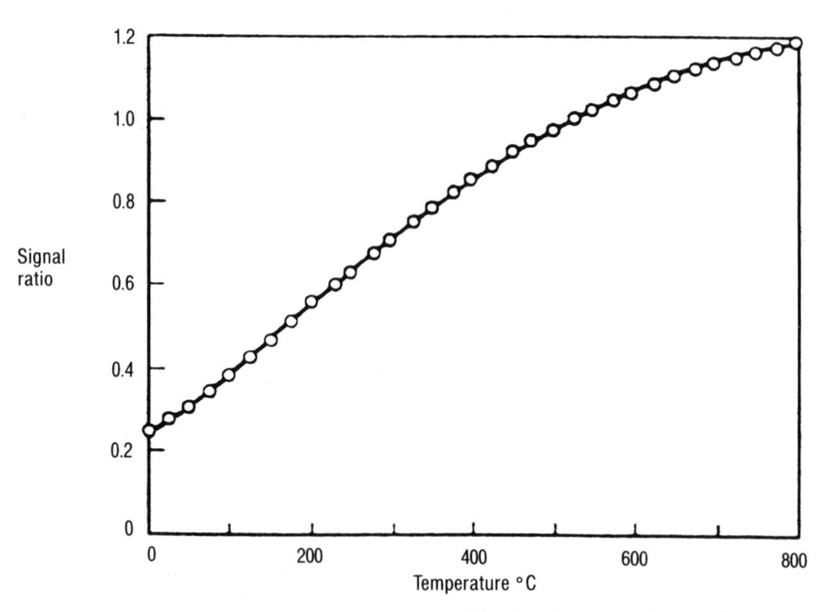

(b) Neodymium Sensor Response

Figure 3-7
Intrinsic Temperature Sensor Using Absorption Characteristics of the Fiber

(reprinted by permission from Optical Society of America)

Fluorescence can be achieved by doping the glass with various additives. The sensor can function in two modes. A light source can be used to stimulate fluorescence, which is affected by temperature; or the fiber can be stimulated by outside radiation and the fluorescence detected, which is a measure of the level of incident radiation, as shown in Figure 3-8.

Refractive index changes can vary the amount of received light by effectively changing the numerical aperture of the fiber. Many polymeric coating materials can be made to have index changes with temperature, thus providing a temperature sensor.

Lastly, doping the glass with various rare earth oxides can make the fiber sensitive to magnetic fields. Such fibers in the presence of magnetic fields rotate the polarized light beam in the fiber, causing a partial extinction and a correlation of light intensity with magnetic field. This concept is referred to as a Faraday rotation.

TRANSMISSION AND REFLECTION WITH OTHER OPTIC EFFECTS

Transmissive sensors can enhance their sensitivity further by adding absorption gratings to the fiber face, as shown in Figure 3-9. To go from maximum intensity now requires movement of only one grating spacing instead of one probe diameter; this could increase sensitivity by a factor of five or more.

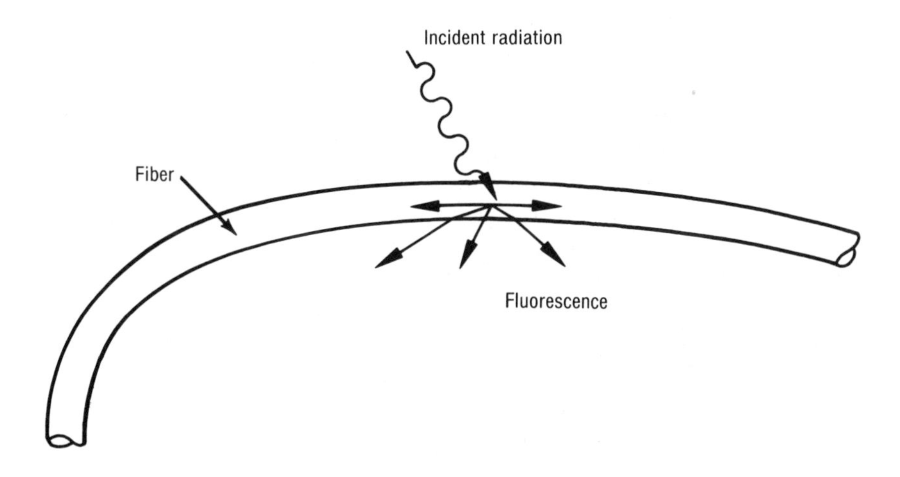

Figure 3-8
Fluorescence Stimulated by Outside Radiation

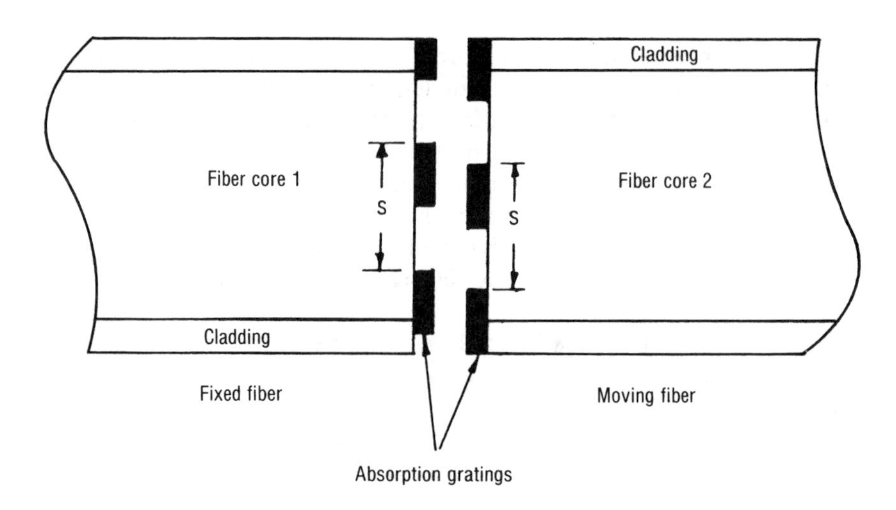

Figure 3-9
Radial Displacement Sensor with Absorption Gratings
(After McMahon, D.H. et al.)
(reprinted by permission from Optical Society of America)

The grating on the fiber is designed to increase sensitivity for radial displacement. The transmissive sensor can be used for rotational sensing. Consider two cases. In case 1, Figure 3-10 depicts two disks: one fixed and the other able to rotate. Each disk has a grating such that the grating of the two disks can lie in an optical line and allow maximum light, or they can be aligned so that there is no light or any position in between. Depending upon the width of the gratings, the sensor changes intensity for a maximum to zero in the rotational distance of 1 grating spacing. The analog signal is linear with the degree of rotation. If two sets of gratings (channels 1 and 2) are used, the direction as well as magnitude can be determined. Since the signal is repeated for each grating spacing, the analog measurement is limited to small rotations such as might be associated with torque.

In case 2, the gratings are replaced with a fixed and a rotating polarizing lens as shown in Figure 3-11 (a). The transmitted intensity through the polarizing lenses is approximately proportional to $\cos^2\theta$[11] where θ is the relative rotation as shown in Figure 3-11 (b). The linear range for this sensor type is limited.

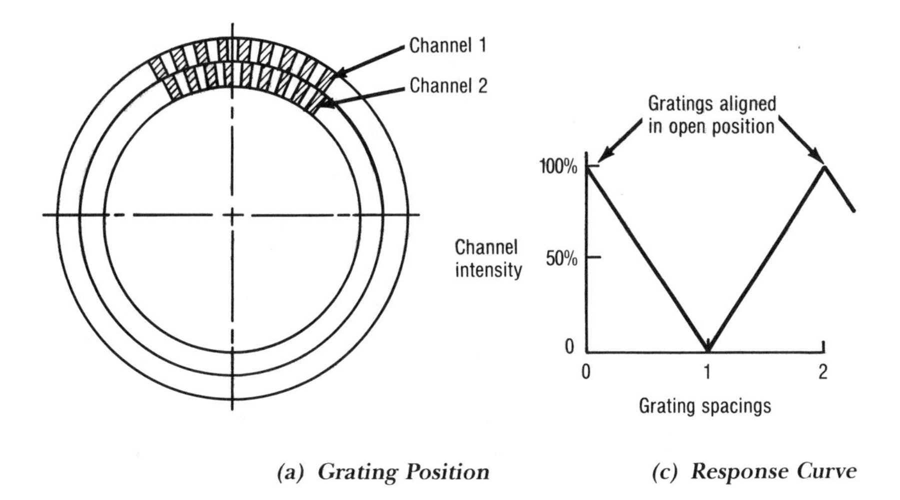

(a) Grating Position *(c) Response Curve*

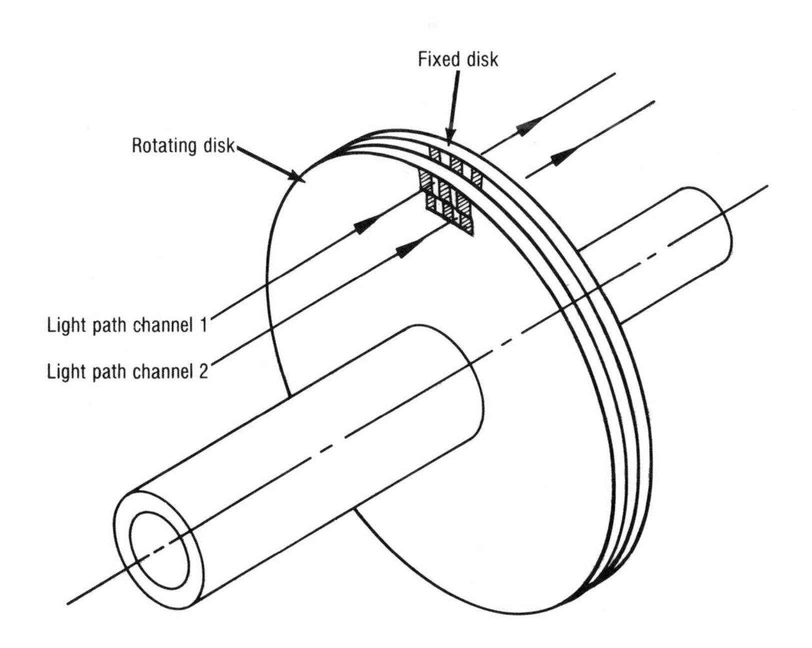

(b) Sensor Configuration

Figure 3-10
Fiber Optic Rotation Sensor

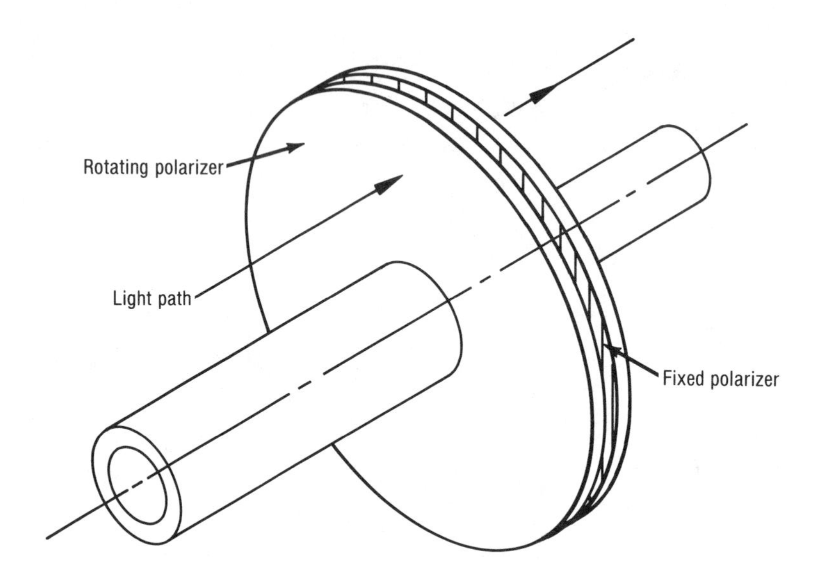

(a) Sensor Configuration

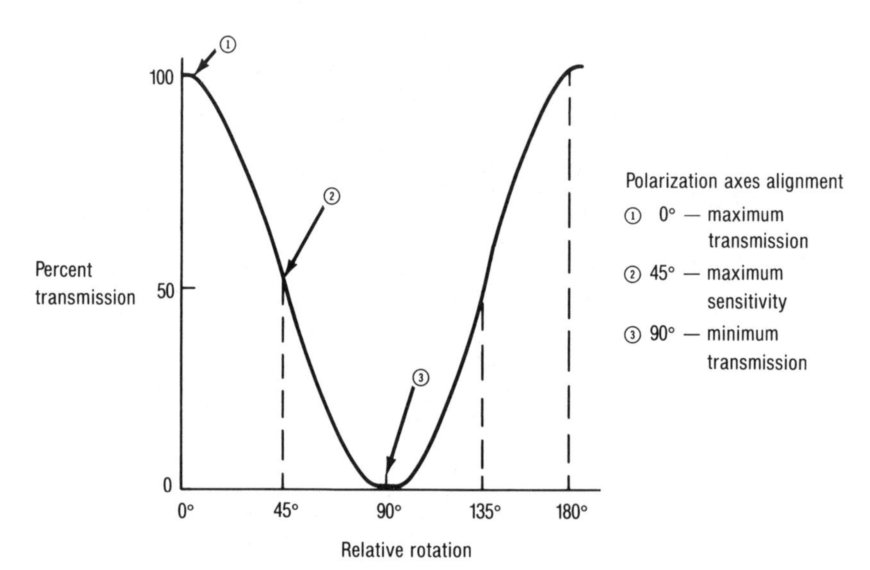

(b) Sensor Response

Figure 3-11
Fiber Optic Rotation Sensor Using Polarization

The transmissive sensor can also be used in conjunction with other optical materials in the light path that alter intensity as a function of environment (see Figure 3-12). If a material in the light path changes transmission levels abruptly with environmental changes, then a switching function is accomplished. Liquid crystals are such materials that can be used for temperature and pressure switching. Analog behavior is observed if the change occurs continuously. A host of materials are available with this behavior. Examples include the following:

(1) Photoelastic materials in conjunction with polarized light are pressure sensitive.
(2) Photochromic material in conjunction with UV light are temperature sensitive.
(3) Materials with dopants providing strong absorption bands are temperature sensitive.
(4) Some photo-luminescent materials alter transmission in the presence of an electric field.

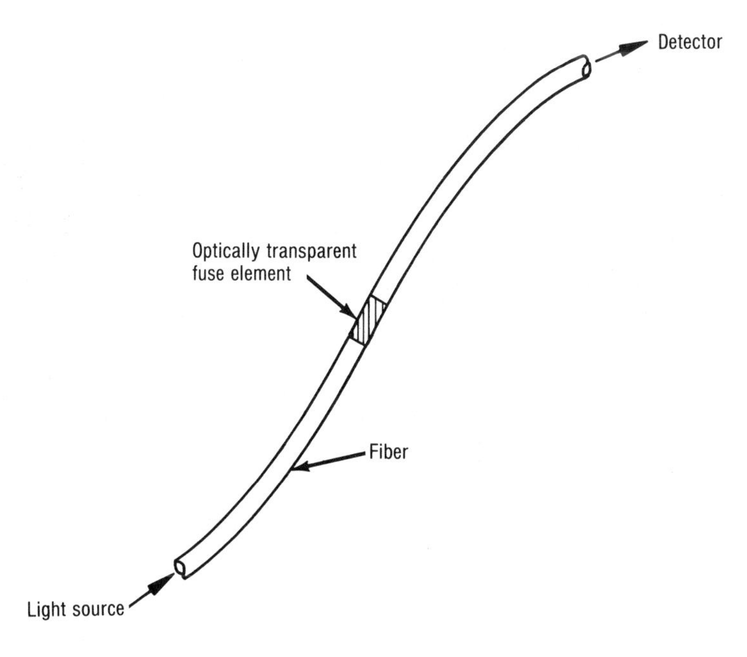

Figure 3-12
Transmissive Sensor with Optically Active Material in Light Path

The list of potential sensing materials is limited only by the imagination of the material scientist.

These active material concepts can be applied to reflective sensors by incorporating the material into the target. An especially attractive reflective sensor concept involves fluorescence. This concept is also referred to as wavelength modulation since the modulated light has a higher wavelength than the incident light. Figure 3-13 depicts how the sensor works. UV light travels down the fiber optic probe and strikes a target, which fluoresces and reflects back into the receiving fiber; often this type of sensor probe uses only a single fiber with transmitted and receiving light rays travelling simultaneously without interfering. The fluorescence is characteristic of the target material. In addition, the intensity of the returning radiation is temperature sensitive. Therefore, the approach has application in both chemical analysis and temperature sensing.

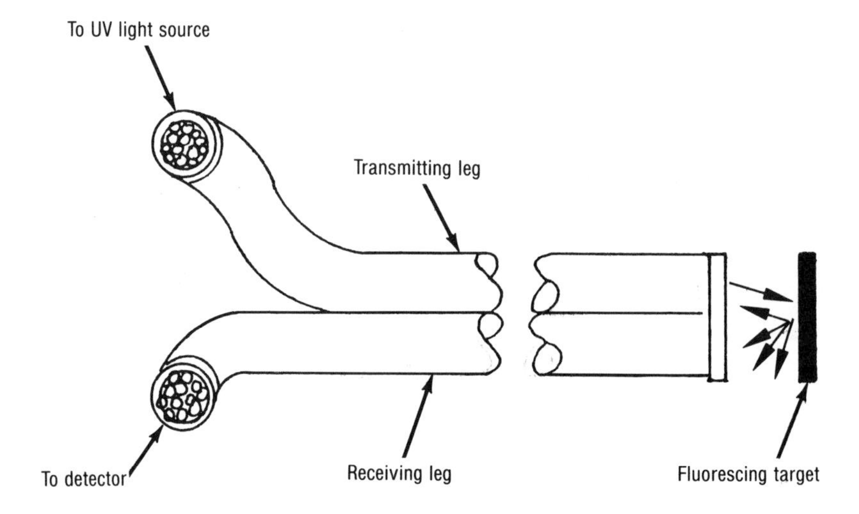

Figure 3-13
Fiber Optic Fluorescence Sensor

REFERENCES

1. Fayfield, R. W., March 1982, "Fiber Optics and Photoelectric Sensing, A Good Combination," *Instruments and Control Systems*, pp 45-49.

2. AMP, 1982, "Fiber Optics Use Gains From Realistic Test Method," *AMP Design Digest*, Vol. 22, p. 6.

3. Spellman, W. B., and McMahon, D. H., 1980, "Frustrated-Total-Internal-Reflection Multimode Fiber Optic Hydrophone," *Appl. Opt.*, Vol. 19, pp 113-117.

4. Kissenger, C. D., and Howland, B., Feb. 24, 1976, "Fiber Optic Displacement Measuring Apparatus," U.S. Patent 3,940,608.

5. Bejczy, A. K., Primus, H. C., and Herman, W. A., March 1980, "Fiber Optic Proximity Sensor," *NASA Tech. Brief*, Vol. 4, No. 3, Item 63, JPL Report NPO-14653/30-4279.

6. Krohn, D. A., 1984, "Fiber Optic Displacement Sensor," *Proceedings of the ISA International Conference — Houston, TX*, Vol. 39, part 1, pp 331-340.

7. McMahon, D. H., Nelson, A. R., and Spillman, Jr., W. B., December 1981, "Fiber Optic Transducers," *IEEE Spectrum*, pp 24-29.

8. Lagakos, N., Litovitz, T., Macedo, P., Mohr, R., and Meister, R., 1981, "Multimode Optical Fiber Displacement Sensor," *Applied Optics*, Vol. 20, No. 2, pp 167-168.

9. Gottlieb, M., and Brandt, G. B., 1979, "Measurement of Temperature with Optical Fibers Using Transmission Intensity Effects," *Proceedings, Electro-Optics Conference*, Anaheim, CA.

10. Snitzer, E., Morey, W. W., and Glenn, W. H., 1983, "Fiber Optic Rare Earth Temperature Sensors," *Proceedings, Optical Fiber Sensors*, p. 79.

11. Anonymous, 1985, *Optics Guide 3*, Melles Griot (Irvine, CA), p. 296.

Phase-Modulated Sensors

4

INTRODUCTION

Phase-modulated sensors are the most publicized of all the fiber optic sensor concepts because of the extreme sensitivity associated with this approach.[8,9] Generally, the sensor employs a coherent laser light source and two single-mode fibers. The light is split and injected into each fiber. If the environment perturbs one fiber relative to the other, a phase shift occurs that can be detected very precisely.[1] The phase shift is detected by an interferometer. There are four interferometric configurations.[2] They include the Mach-Zehnder, the Michelson, the Fabry-Perot, and the Sagnac. The Mach-Zehnder and the Sagnac are the most widely used for hydrophone and gyroscope applications, respectively. However, all four configurations will be discussed in detail in this chapter.

INTERFEROMETERS

Mach-Zehnder

The Mach-Zehnder interferometer configuration is shown in Figure 4-1. The laser output beam is split by using a 3 dB fiber-to-fiber coupler, i.e., 50% of the light is injected into the single-mode sensing fiber and 50% into the reference fiber. The light beams are recombined by using a second 3 dB fiber-to-fiber coupler. The combined beam is detected and the phase shift

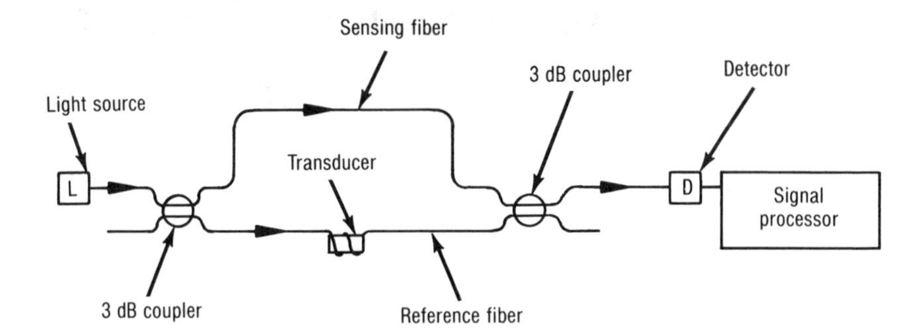

Figure 4-1
Mach-Zehnder Interferometer Configuration
(reprinted by permission from Dynamic Systems, Inc.)

measured. The phase shift results from changes in the length and the refractive index of the sensing fiber. If the path length of the sensing and reference fibers are exactly the same length or differ by an integral number of wavelengths, the recombined beams are exactly in phase and the beam intensity is at its maximum. However, if the two beams are one half wavelength out of phase, the recombined beam is at its minimum value. A modulation of 100% occurs over ½ wavelength of light change in fiber length. This sensitivity allows movements as small as 10^{-13} meters to be detected.

Michelson

The Michelson interferometer configuration is shown in Figure 4-2. The configuration is similar to the Mach-Zehnder approach but uses back reflection caused by the fibers having mirrored ends. The initial coherent laser beam is split and injected into the sensing fiber by the 3 dB coupler. The reference fiber and the sensing fiber have mirrored ends to reflect the beam back through the two fibers and the 3 dB coupler to a detector. The phase shift is then detected. For the Michelson interferometer, a path length difference of $\frac{1}{4}\lambda$ in fiber length results in a $\frac{1}{2}\lambda$ path length change due to the second pass of the reflected beam. The comparison of the Mach-Zehnder and the Michelson interferometers is somewhat analogous to the comparison of transmissive and reflective intensity-modulated sensors. The Michelson approach has the advantage of eliminating one of the 3 dB couplers. However, it has the major disadvantage that the coupler feeds light into both the detector and the laser. Feedback into the laser is a source of noise, especially in high performance systems, and will be discussed in the section on interferometer noise.

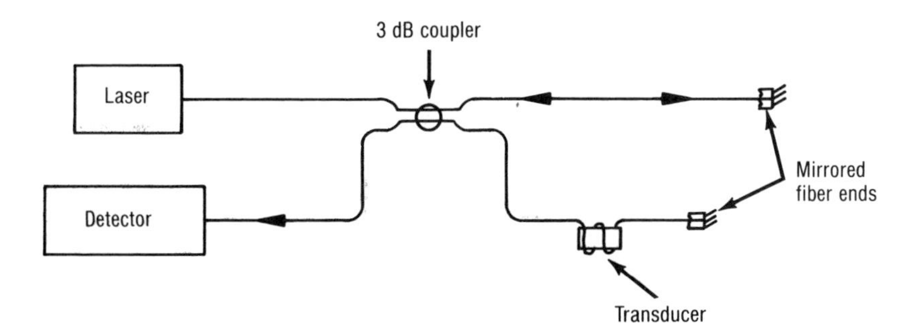

Figure 4-2
Michelson Interferometer Configuration
(reprinted by permission from Dynamic Systems, Inc.)

Fabry-Perot

The Fabry-Perot interferometer involves the concepts previously discussed but does not require a reference fiber. The interference results from successive reflections of the initial beam. The configuration is shown in Figure 4-3. The injected coherent beam is partially reflected back to the laser (typically 95% reflected, 5% transmitted). The transmitted beam that enters the interferometer cavity is partially reflected (95%) and partially transmitted (5%). The 5% of light transmitted by the first mirror is 95% reflected at the second mirror with 5% of the impinging light passing through to the detector. Successive reflection sequences will reduce the detected beam by approximately 10% (5% lost at each of the two reflections per cycle). The multiple passes along the fiber magnify the phase difference, which results in extremely high sensitivity. Generally the Fabry-Perot sensor has twice the sensitivity of the other techniques discussed.

Sagnac

The Sagnac interferometer configuration is shown in Figure 4-4. The Sagnac approach requires that a 3 dB coupler be used to inject light into two ends of a single-mode fiber in a coiled configuration. The injection of light into the fiber is such that light propagates in both clockwise and counterclockwise directions. In this case, both fibers are sensing fibers. While the coil is held stationary, no phase shift occurs since the clockwise and counterclockwise distances are identical. However, if the coil is rotated in one

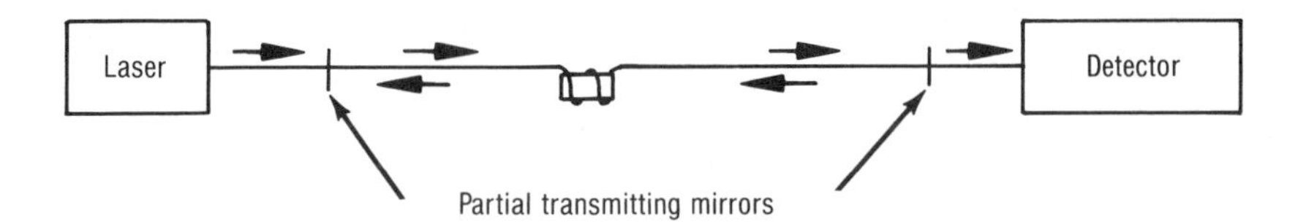

Figure 4-3
Fabry-Perot Interferometer Configuration
(reprinted by permission of Dynamic Systems, Inc.)

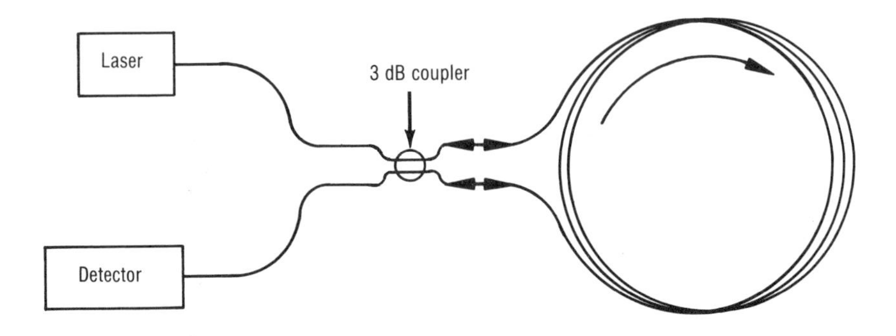

Figure 4-4
Sagnac Interferometer Configuration
(reprinted by permission of Dynamic Systems, Inc.)

direction (for instance, clockwise), the light propagation time is shortened for clockwise propagation. On the other hand, the propagation time is longer for the counterclockwise direction since the path is lengthened. The two recombined beams are then out of phase and a very sensitive rotation measurement is achieved. This approach does not require fiber length and index changes.

PHASE DETECTION

The phase angle, ϕ, for a lightwave travelling in a fiber is defined in Figure 4-5. The phase angle for a lightwave with a given wavelength, λ, and length, L, is given by:

$$\phi = 2\pi L/\lambda = 2\pi n_1 L/\lambda_0 \qquad (4\text{-}1)$$

where n_1 is the index of refraction of the fiber core and λ_0 is the wavelength of light in vacuum. If L and λ_0 are in the same units, the phase angle is in radians. As indicated previously, a change in length and/or refractive index will cause a phase change as defined by the following equation:

$$\phi + \Delta\phi = \frac{2\pi}{\lambda_0}\,[n_1 L + n_1\,\Delta L + L\Delta n_1] \tag{4-2}$$

As a simplification, consider a phase change associated with changes in length. Equation 4-2 simplifies to:

$$\phi + \Delta\phi = \frac{2\pi}{\lambda_0}\,[n_1 L + n_1\,\Delta L] \tag{4-3}$$

Figure 4-5 shows the phase change associated with an increase in length. Consider the case of the Mach-Zehnder interferometer. It consists of 2 fibers: a reference fiber and a sensing fiber. If the sensing fiber is unperturbed, then two fibers have the same length, L. The outputs of the two fibers interfere constructively and give the maximum intensity output. If the sensing fiber

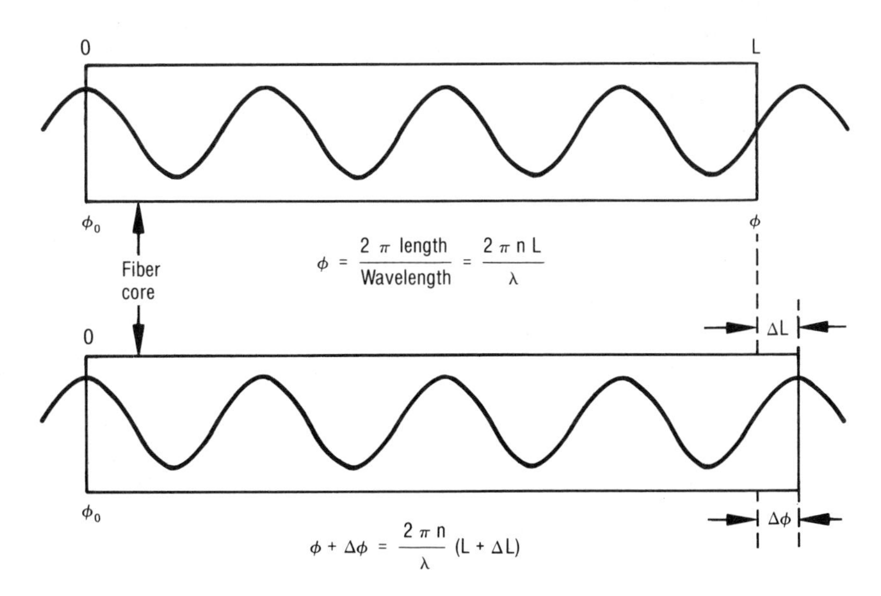

Figure 4-5
Phase Change of a Lightwave through an Optical Fiber of Original Length L That Has Been Stretched by a Length ΔL

(reprinted by permission from Dynamic Systems, Inc.)

experiences a mechanically or thermally applied strain, the sensing fiber increases in length by ΔL. The intensity output decreases due to destructive interference. The relative phase shift versus intensity is shown in Figure 4-6.

The optical intensity is a function of the relative phase shift for the various interferometric configurations. The Michelson and the Sagnac interferometers are similar in intensity output to the Mach-Zehnder interferometer shown in Figure 4-6. When the phase shifts an integral number of wavelengths ($\phi = 0, 2\pi, 4\pi \ldots$), the two legs of the interferometer are in phase, providing constructive interference and maximum intensity. The two light beams destructively interfere and have minimum intensity at a phase shift of an integral number of half wavelengths ($\phi = \pi, 3\pi \ldots$). At the maximum and minimum points, the sensitivity approaches zero. However, in the region of a ¼ wavelength shift ($\pi/2, 3\pi/2 \ldots$) the rate of change of intensity with phase shift is greatest, providing the highest sensitivity, as shown in Figure 4-7.

As stated previously, the Fabry-Perot interferometer is even more sensitive. The intensity versus relative phase shift curve is shown in Figure 4-8. The slope of the curve at the inflective points ($\pi/4, 7\pi/4$) is a measure of the maximum sensitivity, which is twice the value of the other configurations.

As an example of sensitivity, consider the operation of a Mach-Zehnder interferometer at 820 nm using 1 meter of fiber. If the sensor has a resolution of 0.1% and a full scale change is $\lambda/2$, the displacement sensitivity is 0.001 $[820 \times 10^{-9}] = 4.1 \times 10^{-10}$ meters. If 1 km of fiber is used and experiences

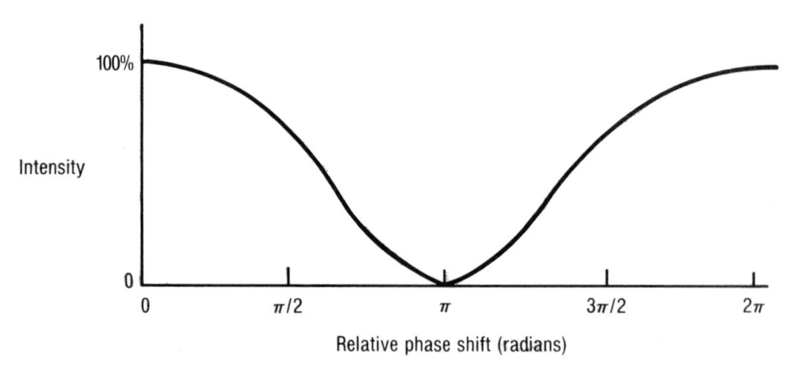

Figure 4-6
Intensity versus Relative Phase Shifts

(reprinted by permission from Dynamic Systems, Inc.)

the same perturbations, then the fiber would have the same phase shift with 1/1000 of the disturbance. The displacement sensitivity is then 4.1×10^{-13} meters, approximately what has been reported as the ultimate sensitivity from the various literature sources.

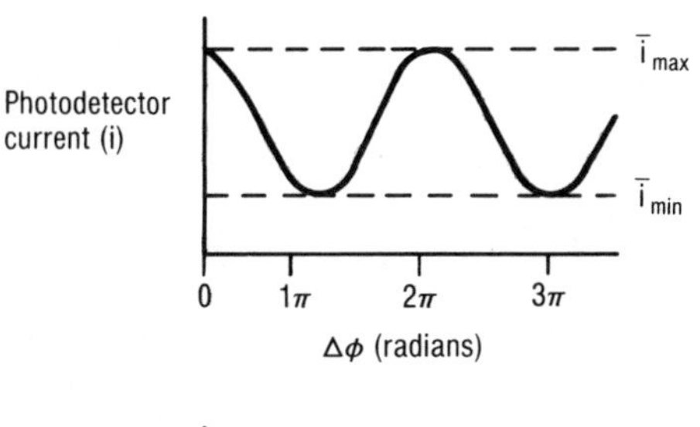

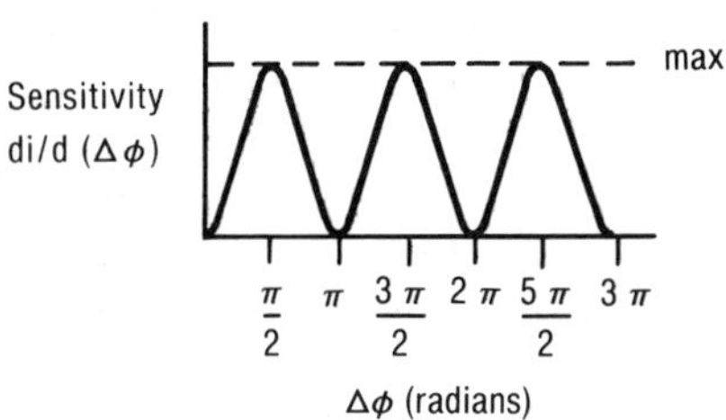

- Phase drift shift causes photodetector current ($\bar{i}$) to vary between $\bar{i}_{min}$ and $\bar{i}_{max}$ (fading)
- Phase sensitivity ~ di/d ($\Delta\phi$)
- Maximum sensitivity for $\Delta\phi = \pi/2$, $3\pi/2$...
- $\pi/2$ phase bias required for maximum sensitivity (quadrature condition)

Figure 4-7
Photodetector Output Current and Its Derivative Resulting from Lightwave Phase Change Fiber Optic Sensor Output

(reprinted by permission from Dynamic Systems, Inc.)

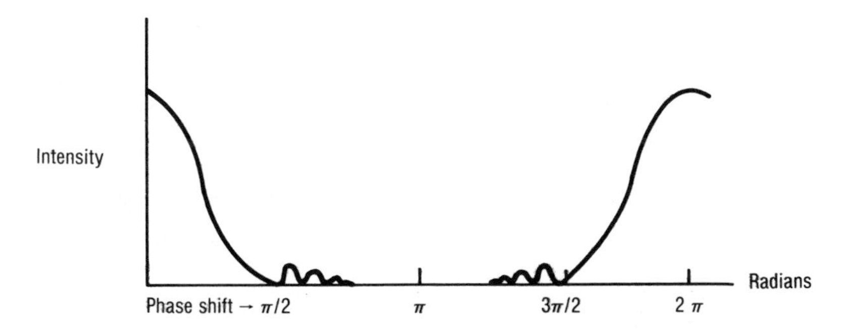

Figure 4-8
Sensitivity versus Phase Shift

(reprinted by permission from Dynamic Systems, Inc.)

The ability to measure displacements at a subatomic level creates a conceptual problem. The sensors are functioning in the world of continuum mechanics, but these displacements are at quantum mechanical levels. It would be difficult to resolve a diaphragm movement, which is 1/2000 of the spacing between atoms comprising the material. The conflict is resolved by considering that interferometers do not directly resolve a mechanical target's movement, but measure the difference between two relatively large values, the difference being quite small.

DETECTION SCHEMES

The Mach-Zehnder interferometer is the most widely used for acoustic sensing; as a result, the detection schemes associated with this configuration will be discussed in some detail. Figure 4-9 shows a Mach-Zehnder interferometer using a homodyne detection scheme. The approach converts phase modulation into intensity modulation.

As discussed previously, the highest sensitivity occurs at a phase shift of 90° with a minimum at 0°. If phase shift drifts from 90° towards 0°, the sensitivity is significantly degraded and causes a condition known as fading. Using a modulator that shifts the phase to a bias condition of 90° creates a condition known as quadrature. The detection scheme is called homodyne detection.

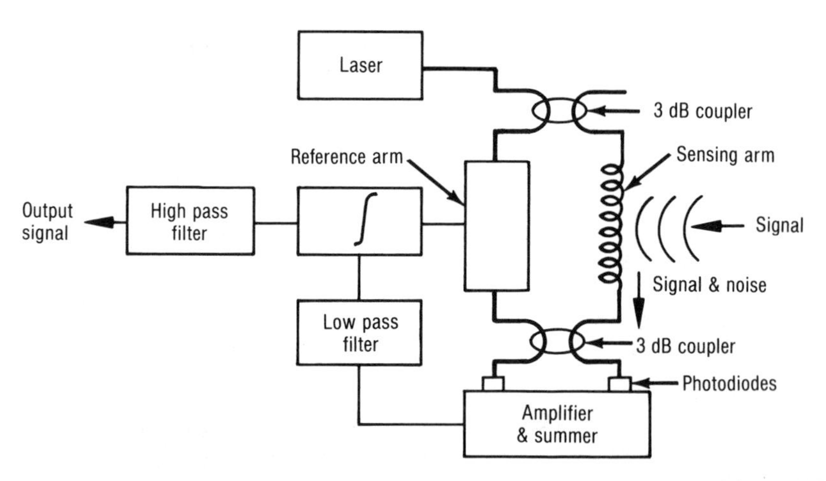

Figure 4-9
A Mach-Zehnder Fiber Optic Interferometer Employing Phase-Locked Homodyne Detection

(reprinted by permission from Dynamic Systems, Inc.)

Referring to Figure 4-9, the laser light is injected equally into the two fibers. The reference fiber and the sensing fiber are clearly marked. The reference arm has a phase shifter to achieve quadrature. The two beams are recombined and detected. The beams are summed, amplified, and fed to a compensation circuit that provides the phase shift signals.

There are several sources of noise in such a configuration, such as phase noise, amplitude noise, and multimode and satellite-mode noise.

Consider phase noise first. The minimal detectable phase shift is a function of the fiber length difference between the two arms of the interferometer. Figure 4-10 is a plot of path length difference versus noise output for various frequencies. As the difference in path length decreases, so does the noise. Figure 4-10 also indicates the minimal detectable phase shift. For the 2 kHz curve, if the two interferometric fibers are matched to within 1 mm, phase shifts of 10^{-6} radians are detectable. Matching the two legs to within 1 mm is the practical limit due to mechanical and thermal strains. Typical fibers are 100 meters or more long. Matching to 1 mm gives a length accuracy of 0.001%. If the same fibers are matched only to within 1 meter

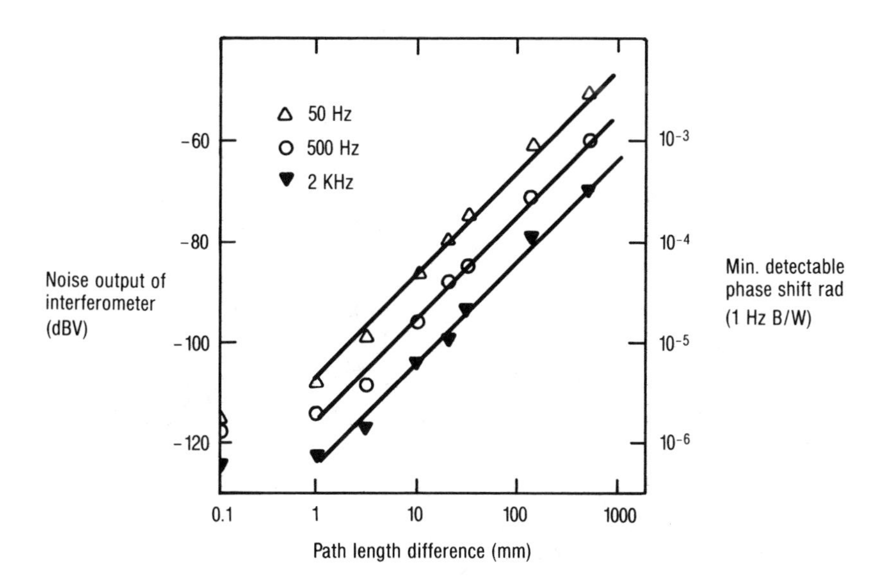

Figure 4-10
Variation of Homodyne Interferometer Output Noise as a Function of Sensing Arm Path Length Difference for Several Output Frequencies

(reprinted by permission from American Institute of Physics)

(1%), the detectable phase shift increases to 10^{-3} radians — three orders of magnitude less sensitive. Phase noise is also due to the laser source and, to a lesser extent, the photodiode response.

Amplitude noise results from fluctuations in the output intensity of the laser source. By using summing techniques in the electronics, the variation can be mathematically eliminated. The technique is called common mode rejection. Using the technique at low frequencies, the minimum detected phase shift is improved by an order of magnitude; at higher frequency the advantage is not as pronounced.[2]

Multimode operation in the fiber creates undesirable interference and limits phase shift detection. Back reflections from the fiber/coupler interface cause multimodes to form. The effect of increasing back reflection is shown in Figure 4-11. To minimize the effect, Fresnel reflections at the fiber/

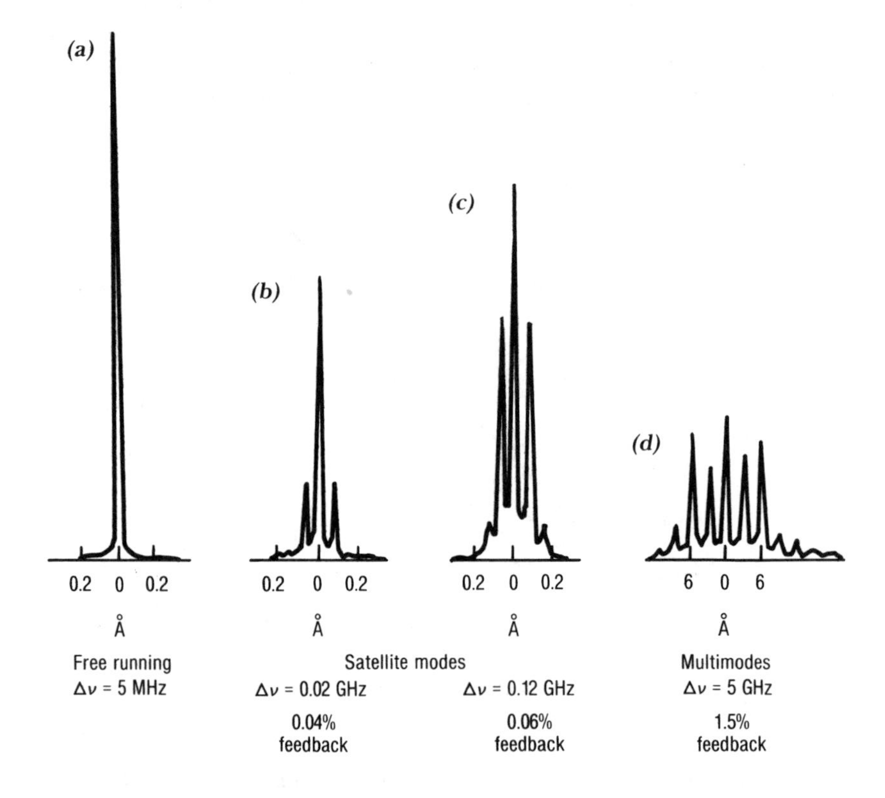

Figure 4-11
Influence of Optical Feedback on the Modal Output of a Diode Laser

(reprinted by permission from Dynamic Systems, Inc.)

coupling interfaces have to be reduced. This can be achieved by careful fiber alignment and index-matching fluids at the interface. A new technology is evolving in which the fibers are polished with a slightly convex surface, which nearly eliminates back scatter.

A heterodyne detection scheme is shown in Figure 4-12. The device employs Bragg modulators. The beam is split at a 3 dB coupler and enters the two Bragg modulators, which shift the wavelength. The frequency of the two modulators is in the range of 50 MHz with a difference between the sensing leg and reference leg of approximately 100 kHz. This frequency shift permits the device to work in the sensitive phase shift regions without the feedback circuiting required for quadrature. Since the device is less sensitive to low frequency noise and optical intensity fluctuations, process circuitry is further simplified.

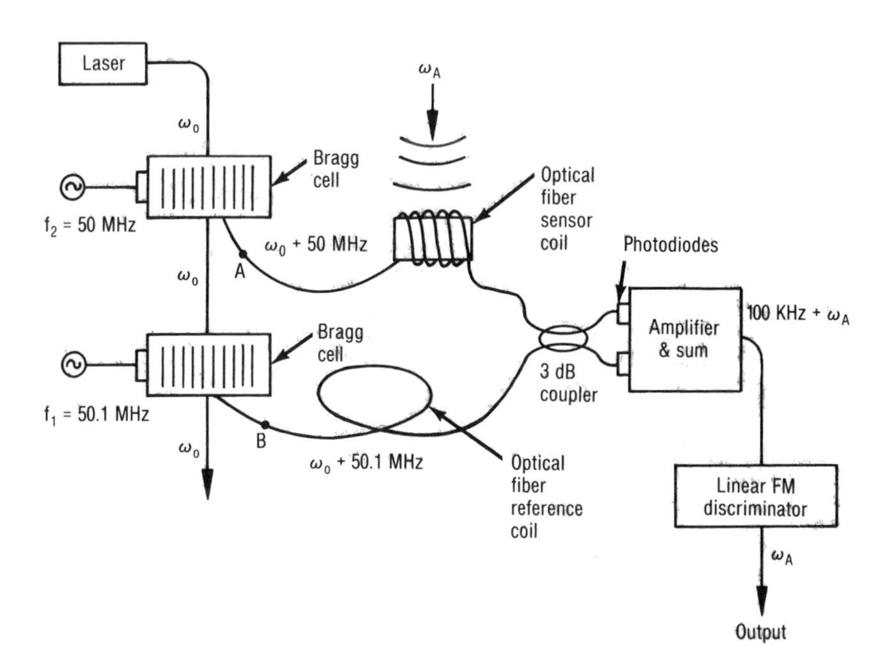

Figure 4-12
An Interferometric Fiber Optic Sensor Employing Heterodyne Detection

(reprinted by permission from Dynamic Systems, Inc.)

POLARIZING MAINTAINING FIBERS

In the case of interferometric fiber optic sensors, the description assumes a single-mode fiber in which phase shifts between a sensing and a reference fiber can be determined. In reality, single-mode fibers are really dual-mode fibers due to the fact that there are two possible degenerate polarization modes.[7] The fundamental HE_{11} mode can be separated into the horizontal or H-mode and the vertical or V-mode as shown in Figure 4-13. To achieve the desired sensing properties, preservation of polarization in single-mode fibers is required. Two approaches can be used: a circumferential stress, due to an expansion mismatch and/or a noncircular core, eliminates the degeneracy. These fiber configurations are shown in Figure 4-14.

The modes in the vertical and horizontal directions propagate at different phase velocities in polarizing maintaining fibers and as such have high effective indices of refraction in these directions. The difference between the indices defines model birefringence, *B*.

$$B = n_V - n_H \tag{4-4}$$

where n_V and n_H are the indices in the vertical and horizontal directions, respectively.

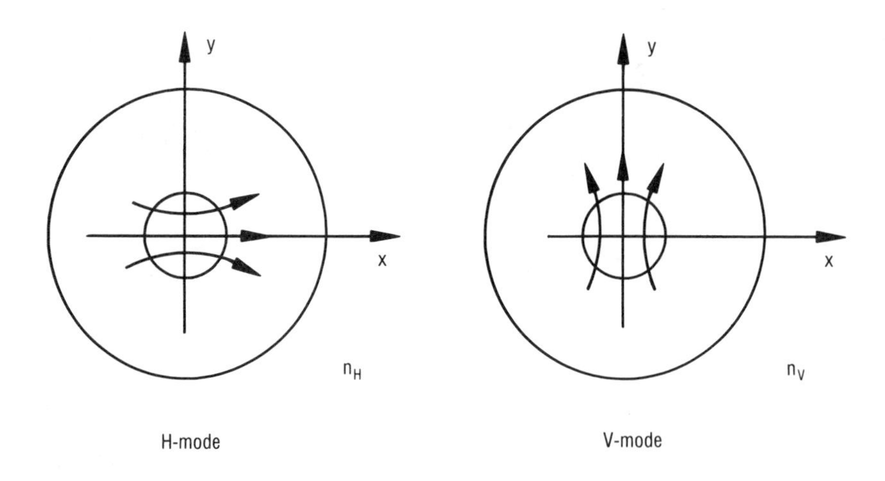

Figure 4-13
The Two Polarizations of the Fundamental HE_{11} Mode in a Single-Mode Fiber
(reprinted by permission from the Optical Society of America)

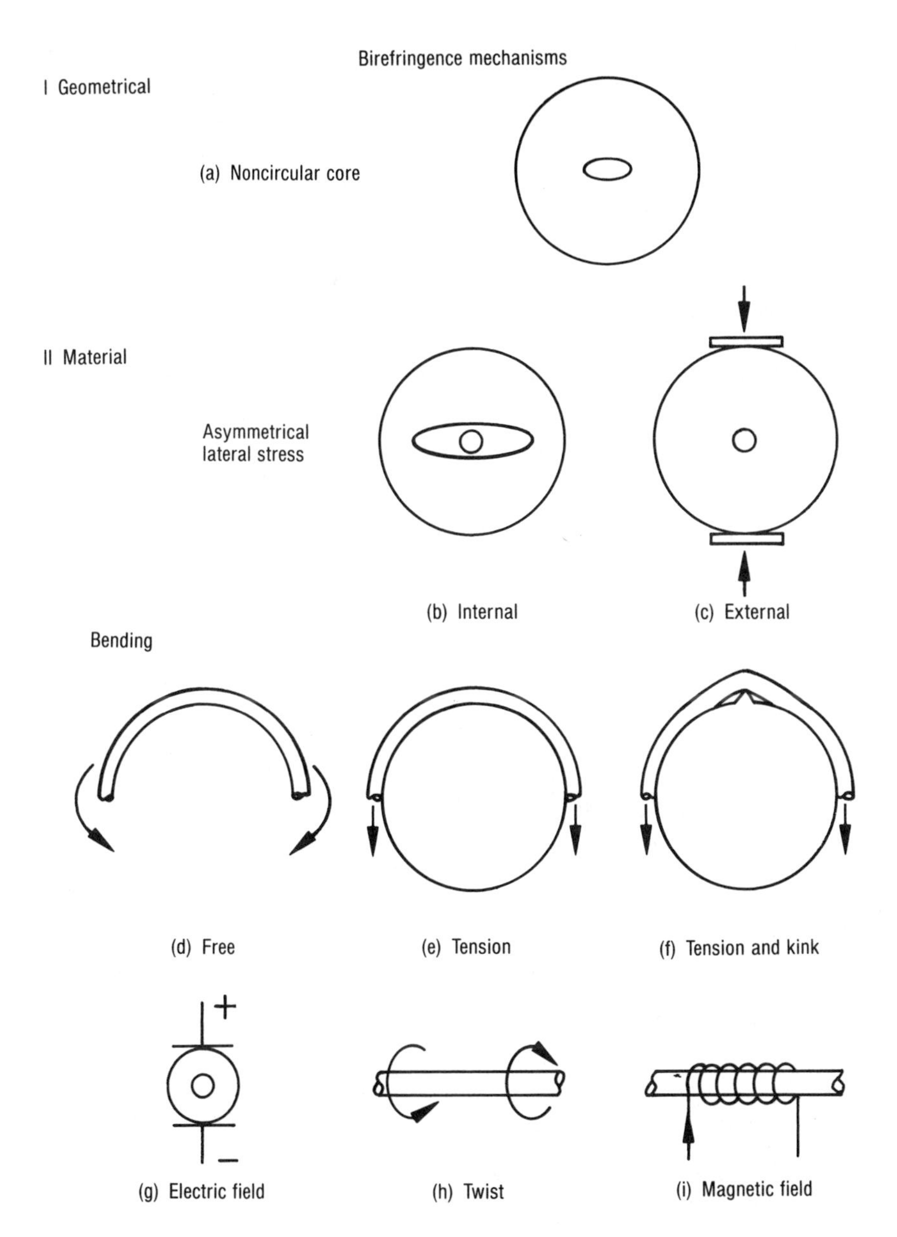

Figure 4-14
Birefringence from Waveguide Geometry and from Various Internal and External Elasto-optic, Magneto-optic, and Electro-optic Contributions

(reprinted by permission from the Optical Society of America)

The effect of birefringence on the guided light is shown in Figure 4-15. Light linearly polarized at 45° to the horizontal and vertical directions is injected into the fiber. The polarization state changes continuously along the fiber due to the fact that as the light propagates, the *V*-mode slips in phase relative to the *H*-mode. The linear polarization goes to elliptically polarized light and then circularly polarized light at a phase shift of $\pi/2$. The circularly polarized light goes through the elliptical state and to linear polarization at phase shift of π. However, the polarization vector is now 180° out of phase with the initial polarization state.

Figure 4-16 helps to further clarify the effects of birefringence. If properly positioned at 0° and 2π radians rotation, scattered light is visible through

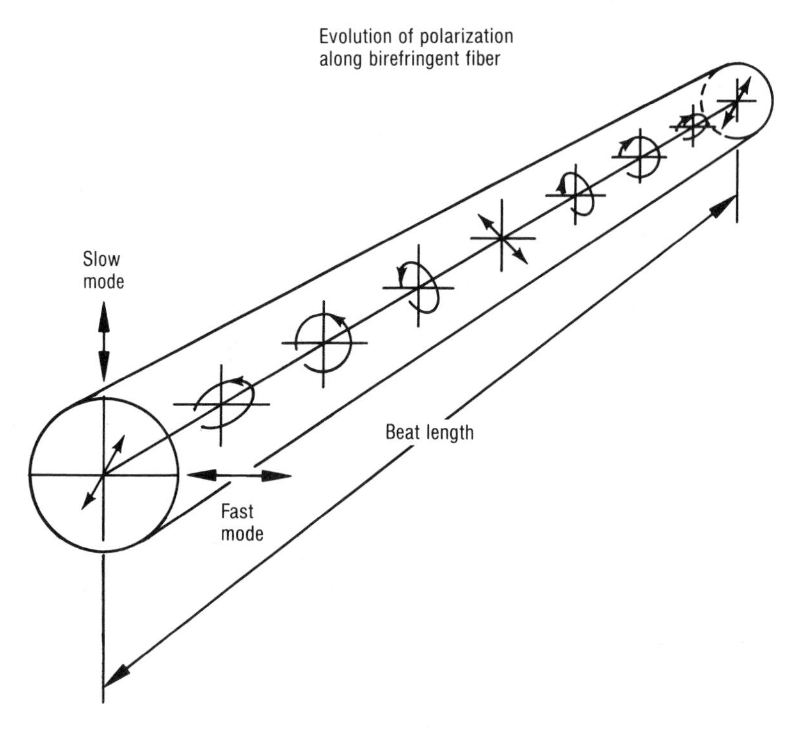

Figure 4-15
The Effect of Uniform Birefringence on the State of Polarization of Initially Linearly Polarized Light Injected at 45° to the Fibers' Principle Axes

(reprinted by permission from the Optical Society of America)

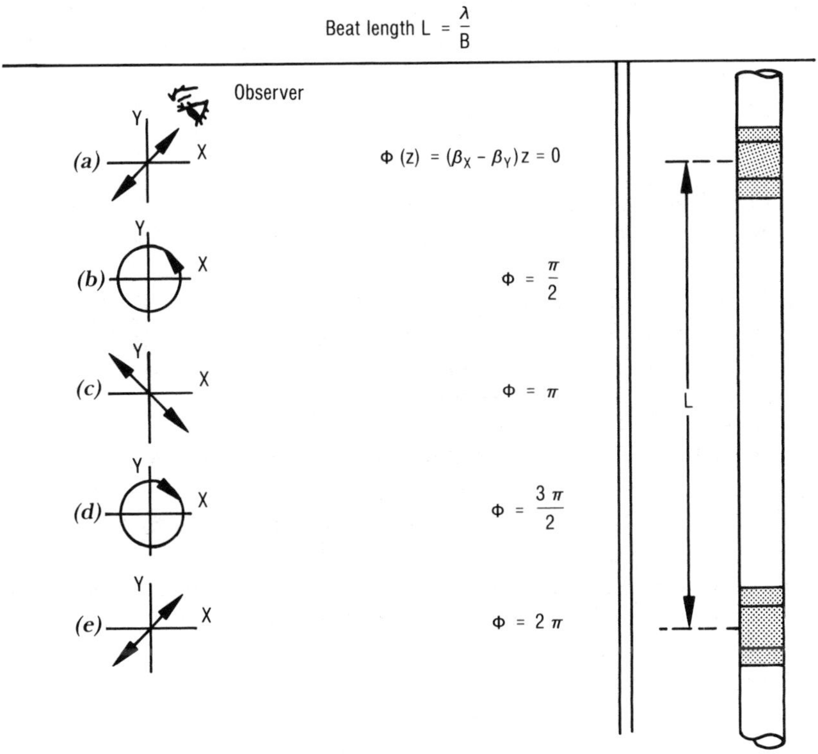

Figure 4-16
The Effects of Birefringence on Polarized Light in Optical Fiber
(© 1981 IEEE; reprinted with permission)

the sides of the fiber. The distance, L, for a complete rotation of the polarization state is defined as the beatlength.*

$$L = \lambda / B \tag{4-5}$$

Typical beat lengths are between 10 cm and 2 m. A more important parameter to charaterize polarization maintaining fiber is the h-parameter,

*The beatlength term here is defined differently from the term described in Chapter 2.

the so-called polarization maintaining property of the fiber. It describes the cross coupling of power injected in the H-mode to the V-mode. The power cross coupled is given by

$$\frac{P_V}{P_V + P_H} = \tfrac{1}{2}\left(1 - e^{-2hL}\right) \tag{4-6}$$

where P_V and P_H are the powers in the vertical and horizontal polarization modes, respectively. It is clear that as the length of the fiber increases, the cross-coupled power increases. To achieve the desired "true" single-mode operation, P_V should approach zero and h should be very small. Typical values of h for good quality polarizing maintaining fibers range for 5×10^{-6}/m to 2×10^{-5}/m. An h-value of 10^{-5}/m results in 1% of cross-coupled power for a fiber 1 kilometer long.

In applications with interferometric sensors, it is necessary that the output beam be matched in not only amplitude but in polarization direction if constructive interference is to occur. Polarization maintaining fibers improve sensitivity using fiber with high birefringence, short beatlength and low h-values. It is important to note that fiber with high birefringence can have structural perturbations that can increase both cross coupling and h-values. The ability of fiber to hold polarization state minimizes the possibility of degenerate modes that degrade sensitivity.

REFERENCES

1. Mellberg, R. S., 1983, "Fiber Optic Sensors," *SRI International*, Research Report No. 684.

2. Davis, C. M., et al., 1982, *Fiber Optic Sensor Technology Handbook*, Dynamic Systems, Reston, Virginia.

3. Rashleigh, S. C., and Stolen, R. H., 1982, "Preservation of Polarization in Single Mode Fibers," Presented at the Annual Meeting of the Optical Society of America.

4. Dandridge, A., Tveten, R., Miles, R., Jackson, D., and Giallorenzi, T., 1981, "Single Mode Diode Laser Phase Noise," *Applied Physics Letters*, Vol. 38, p. 77.

5. Miles, R., Dandridge, A., Tveten, A., Taylor, H., and Giallorenzi, T., 1980, "Feedback Induced Line Broadening in CW Channel-Substrate Laser Diodes," *Applied Physics Letters*, Vol. 37, p. 990.

6. Wahl, J. F., 1985, "Characterization of Special Fibers," *Guidelines*, Corning, NY, p. 3.

7. Stolen, R. H., 1986, "Polarizing — Preserving Fiber Optics," SPIE's —O-E LASE, Tutorial T26.

8. Tebo, A. R., 1982, "Sensing with Optical Fibers: An Emerging Technology," *Proceedings of the ISA International Conference*, Philadelphia, PA, pp 1655–71.

9. McMahon, A. R., Nelson, and Spillman, W. B., 1981, "Fiber Optic Tranducers," *IEEE Spectrum*, pp 24–29.

10. Kaminow, I. P., 1981, "Polarization in Optical Fibers," *IEEE Journal of Quantum Electronics*, Vol. QE-17, No. 1, pp 15–22.

Digital Switches and Counters

5

INTRODUCTION

All intensity-modulated fiber optic sensors are analog in nature, but they have a broad usage potential in digital applications. Until recently they were used as absence/presence sensors, i.e., is the object there or not? Digital applications have now expanded to pressure, temperature, and liquid level switches. Photoelectric sensors have been used for many years in absence/presence applications. Since fiber optic sensors are almost always more expensive than photoelectric sensors, the driving force for using fiber optics is generally due to environmental considerations. Fiber optic sensors, due to their small mass, are less susceptible to vibration. Fiber optic devices are immune to electromagnetic interference and radio frequency interference and, therefore, are free from electrical noise problems. They are explosion-proof. Their small size makes them easy to install, and conduit and shielding are generally not necessary. Figure 5-1 shows photographs of photoelectric and fiber optic sensors for comparison.

SCAN MODES

The two basic scan modes are through-scan and reflective scanning. Through-scan uses fibers in an opposed transmissive mode, as shown in Figure 5-2. The transmitting fiber optic bundle is opposed by the receiving bundle. If an object interrupts the light path, the object is detected. The bundles can be replaced by large-core single fibers, but bundles provide more light and a larger scan area than single fibers. Therefore, the choice of bundle versus single fiber depends upon the application. The through-scan approach provides for the maximum collection of light; therefore, it is the most effective technique in dirty environments and for long distances between the sensor and the object. It is especially well suited for detecting opaque objects, but sufficient light may penetrate translucent objects to prevent detection.

Reflective scanning uses fiber optic probes that are bifurcated, i.e., the transmit and receive legs are separate at the electro-optic interface and have

Figure 5-1
Photoelectric and Fiber Optic Sensors

(courtesy of Micro Switch)

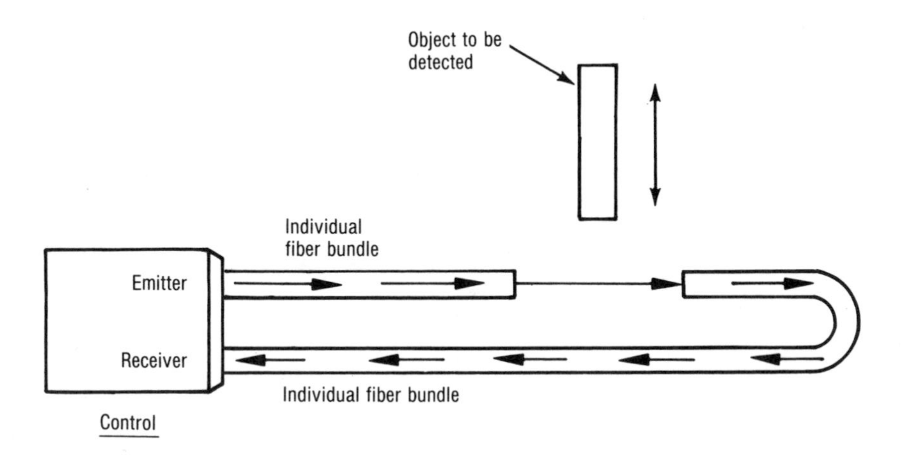

Figure 5-2
Through-Scan Mode
(courtesy Banner Engineering Corp.)

a common leg near the target. Light is transmitted to the target through the transmitting portion of the common leg, reflected from the target, and returned to the photodetector via the receive portion of the common bundle. An object is detected when it interrupts the light path between the target and the probe. The technique works well for object detection in which surface reflectivity is poor. If a highly reflective object breaks the light beam, there may be sufficient reflected light collected by the receiving leg to generate a false signal. Reflective scanning is characterized by the target reflectivity, i.e., retro-reflective, diffuse, and specular.

A retro-reflective fiber optic sensor is shown in Figure 5-3. The retro-reflecting target is highly efficient in reflecting light back along the same light path as the incident beam. The reflection phenomenon is independent of the angle at which the light initially hits the retro-reflector. Figure 5-4 shows the topography of retro-reflective surfaces that allow for high reflection efficiency. The reflectors use either spherical balls or 3-corner cubes.

The 3-corner cube also has the unique property of being a polarizing retarder. In essence, if polarizing filters with their axes at 90° to each other are placed at the receiving and transmitting ends of the probe, the plane of polarization is rotated effectively 90°. Only light reflected from the 3-corner cube can pass through the receiving polarizing filter. All other light is rejected. This feature eliminates false signals from highly reflective objects, as discussed earlier.[3]

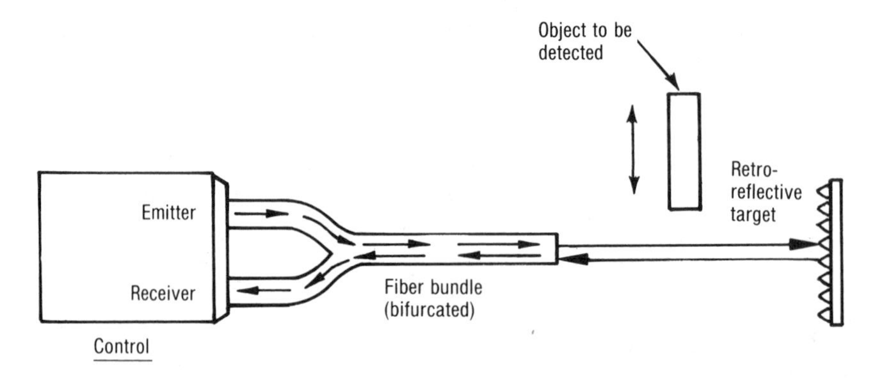

Figure 5-3
Retro-Reflective Fiber Optic Sensor
(courtesy Banner Engineering Corp.)

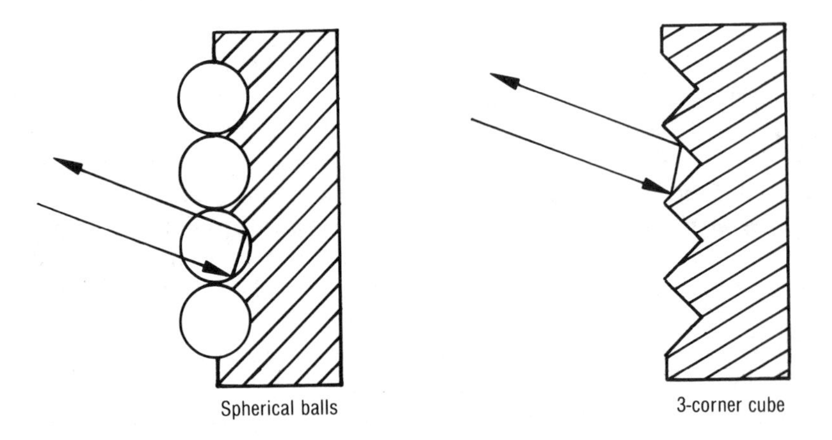

Figure 5-4
Retro-Reflective Surface Topographies

Diffuse scanning occurs for targets with relatively rough sufaces. The initial incident light beam is scattered in numerous directions with the received light level being low. In this case, the reflecting target is the object to be detected, as shown in Figure 5-5. The primary application is in proximity sensing.

A potential problem can occur if the background has a high reflectivity relative to the target material. In such cases, false triggering is likely.

Specular targets are the third application of reflective scanning. The surface is characterized as being smooth and shiny or mirror-like. The angle of incidence is equal to the angle of reflection, as shown in Figure 5-6.

For distances that are small (usually less than ¼ inch), bifurcated probes can be used. However, for larger distances between the target and the probe or for large angles of incidence, the transmit and receive legs require separation, as would be the case for a transmissive through-scan mode. This technique is especially useful in differentiating between highly reflective surfaces and dull areas.

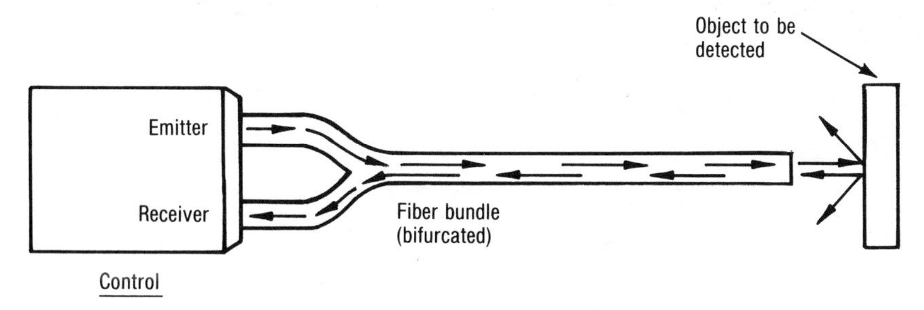

Figure 5-5
Reflective Fiber Optic Diffuse Scanning
(courtesy Banner Engineering Corp.)

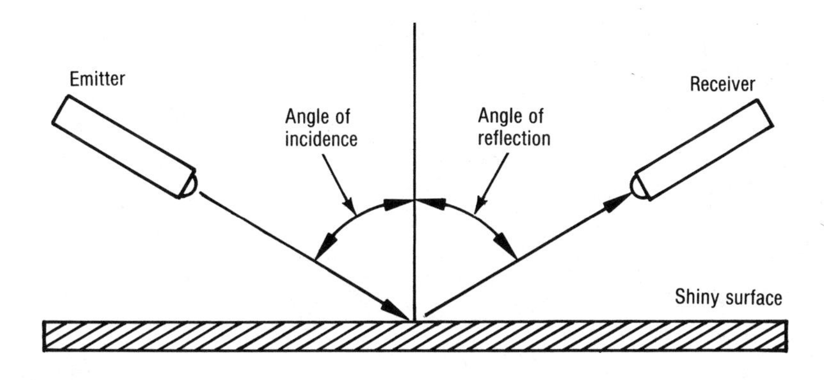

Figure 5-6
Specular Scanning

EXCESS GAIN

Excess gain is one of the parameters used to characterize absence/presence sensors.[2,3] It is defined as the ratio of light intensity hitting the detection system to the light intensity required to activate the detection system. Therefore, a ratio of one is just sufficient to operate the system. Generally, an excess gain is specified to handle degradation of the system caused by airborne contamination, dirty optics, potential misalignment, surface reflectivity changes, etc. Typical excess gain curves for the four modes discussed (through-scan, retro-reflective, diffuse, and specular) are shown in Figures 5-7, 5-8, 5-9, and 5-10, respectively.

Tables 5-1 and 5-2 have been worked out for photoelectric sensors but they will be useful, at least for a relative level of excess gain requirements.

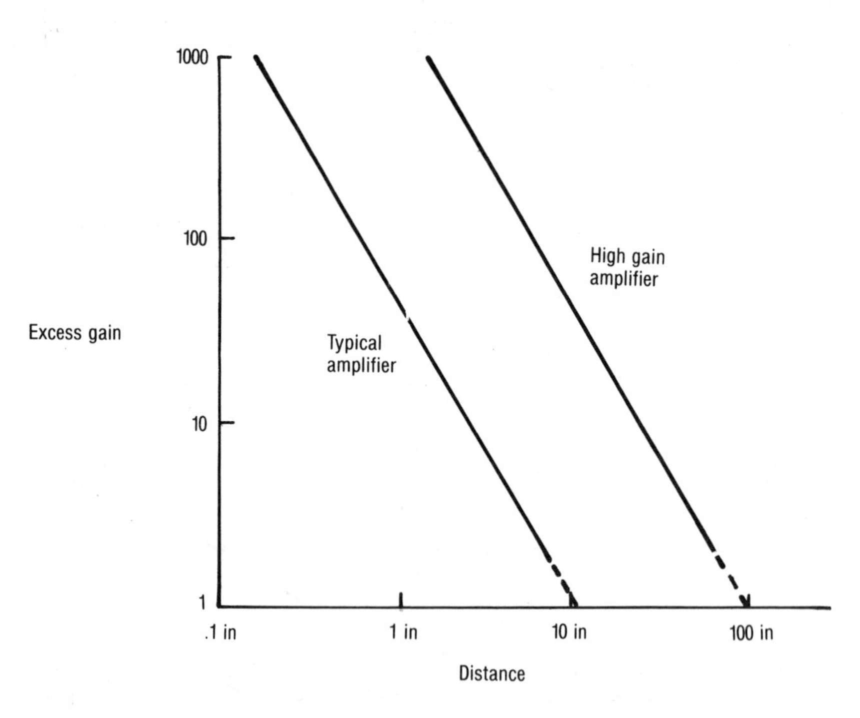

Figure 5-7
Excess Gain Through-Scan
(courtesy Banner Engineering Corp.)

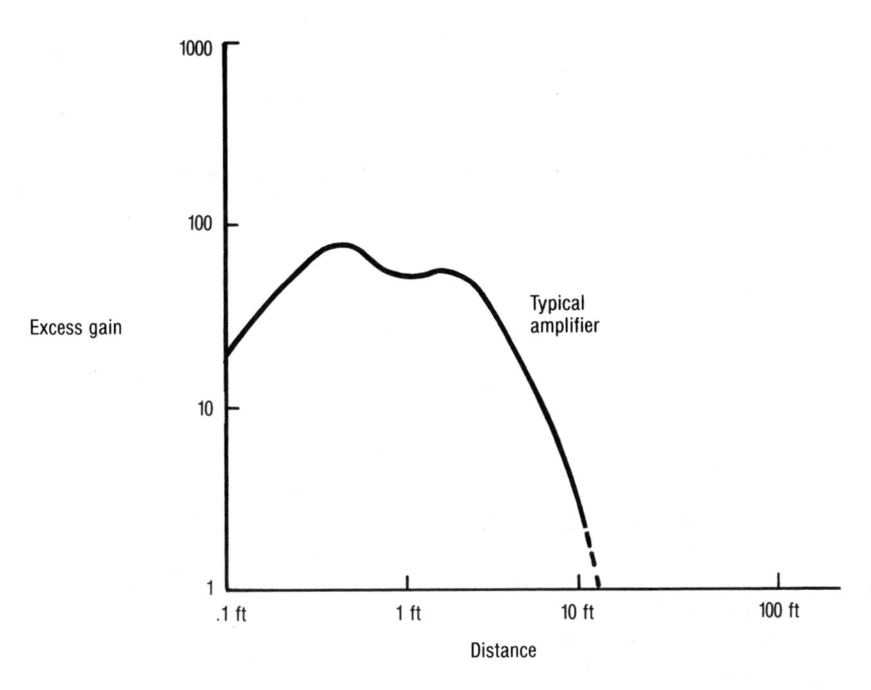

Figure 5-8
Excess Gain Retro-Reflective
(courtesy Banner Engineering Corp.)

Table 5-1
Excess Gain Required for Various Operating Environments

Minimum Excess Gain Required	Operating Environment
1.5X	Clean air: No dirt build-up on lenses or reflectors.
5X	Slightly dirty: Slight build-up of dust, dirt, lint, moisture, oil film, etc. on lenses or reflectors; lenses cleaned regularly.
10X	Moderately dirty: Obvious contamination of lenses or reflectors, but not obscured; lenses cleaned occasionally or when necessary.
50X	Very dirty: Heavy contamination of lenses; heavy fog, mist, dust, smoke, or oil film; minimal cleaning of lenses.

(reprinted by permission from Banner, Inc.)

Table 5-2
Relative Reflectivity Chart

Material	Reflectivity (%)	Excess Gain Required
Kodak white test card	90%	1
White paper	80%	1.1
Newspaper (with print)	55%	1.6
Tissue paper: 2 ply	47%	1.9
1 ply	35%	2.6
Masking tape	75%	1.2
Kraft paper, cardboard	70%	1.3
Dimension lumber (pine, dry, clean)	75%	1.2
Rough wood pallet (clean)	20%	4.5
Beer foam	70%	1.3
*Clear plastic bottle	40%	2.3
*Translucent (brown) plastic bottle	60%	1.5
*Opaque white plastic	87%	1.0
*Opaque black plastic (nylon)	14%	6.4
Black neoprene	4%	22.5
Black foam carpet backing	2%	45
Black rubber tire wall	1.5%	60
*Natural aluminum, unfinished	140%	0.6
*Natural aluminum, straightlined	105%	0.9
*Black anodized aluminum, unfinished	115%	0.8
*Black anodized aluminum, straightlined	50%	1.8
*Stainless steel, microfinish	400%	0.2
*Stainless steel, brushed	120%	0.8

*Note: For materials with shiny or glossy surfaces, the reflectivity figure represents the maximum light return, with the scanner beam exactly perpendicular to the material surface.

(reprinted by permission from Banner, Inc.)

CONTRAST

The use of absense/presence sensors requires that the sensor be able to differentiate between two different light levels. Contrast is defined as the ratio of the level of detected light in the light condition to the level of detected light in the dark condition. Generally, if the contrast ratio is less than 3, the contrast is considered poor, and care must be exercised in choosing a detection scheme. Above a ratio of 3, contrast does not normally present a problem. A common problem mentioned previously is a highly

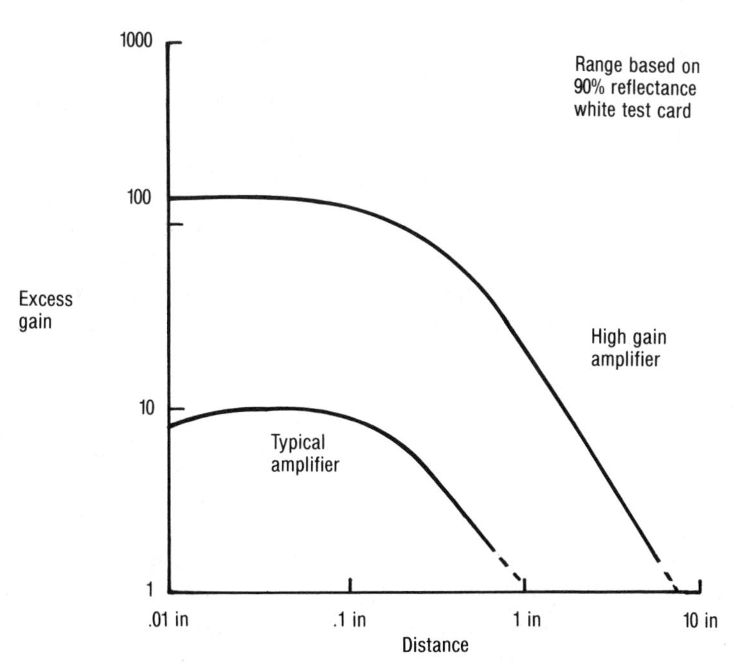

Figure 5-9
Excess Gain Diffuse Scan

(courtesy Banner Engineering Corp.)

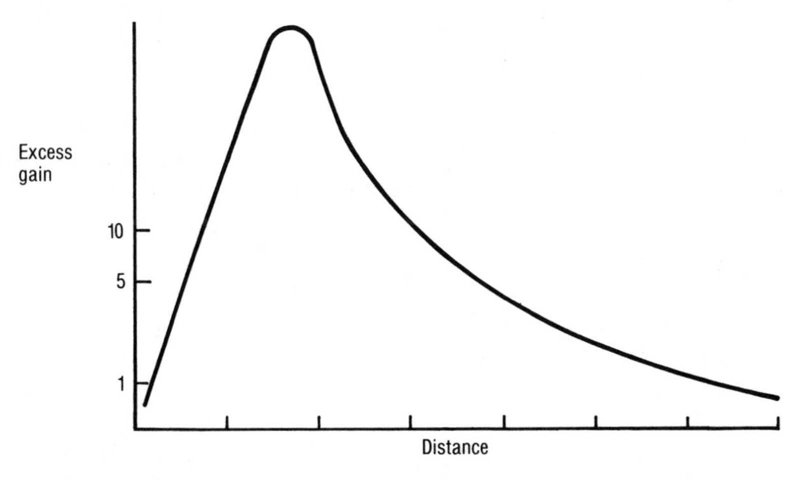

Figure 5-10
Excess Gain Specular

reflective object interrupting the light beam in a diffuse scanning mode. False triggering can be expected. To eliminate this problem, the reflection target can be replaced with a retro-reflective material, or through-scanning can be used. The retro-active target increases the contrast ratio so that the interruption of the light path by a highly reflectiive object still is way below the detected signal from the target. Using a through scan mode eliminates the possibility of detecting reflected light. Another potential problem is a transparent or translucent object breaking the light beam in a retro-reflective mode or a through-scan mode. Sufficient light may pass through the object that a false signal will occur. In this case, the diffuse reflective mode may be best.

BEAM DIAMETER

A second characterization parameter is the beam diameter. It is defined by the locus of points at which the excess gain is equal to 1. It defines the off-axis distance at which the system will still operate. The highest energy is on or near the axis of the transmitting leg, with the transmitted energy falling off as the distance parallel to the axis or perpendicular to the axis increases (see Figure 5-11).

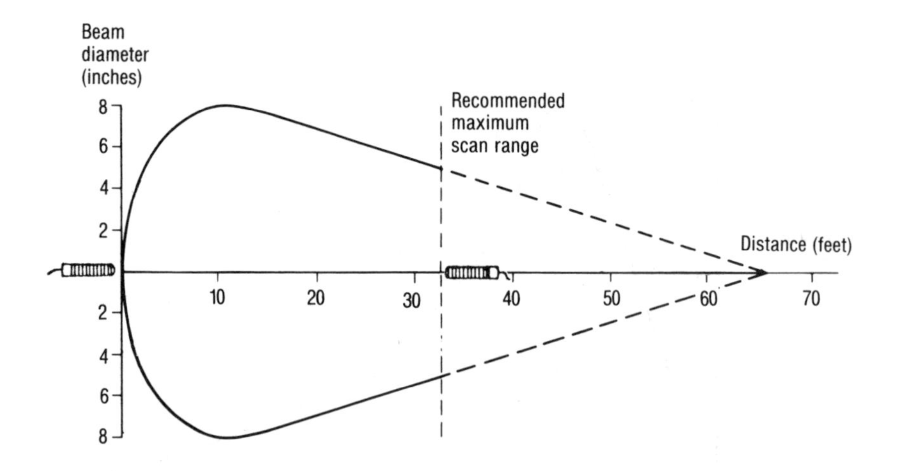

Figure 5-11
Beam Diameter

(reprinted by permission from Veeder-Root)

The effective beam diameter is the portion of the beam diameter that reaches the receiver. For a through-scan configuration, Figure 5-12 defines the effective beam. Figure 5-13 defines the effective beam for a reflective scan.

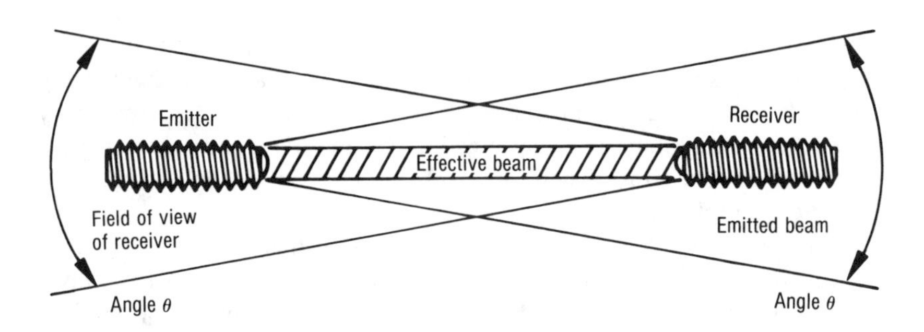

Figure 5-12
Effective Beam Diameter — Through-Scan Configuration
(courtesy Banner Engineering Corp.)

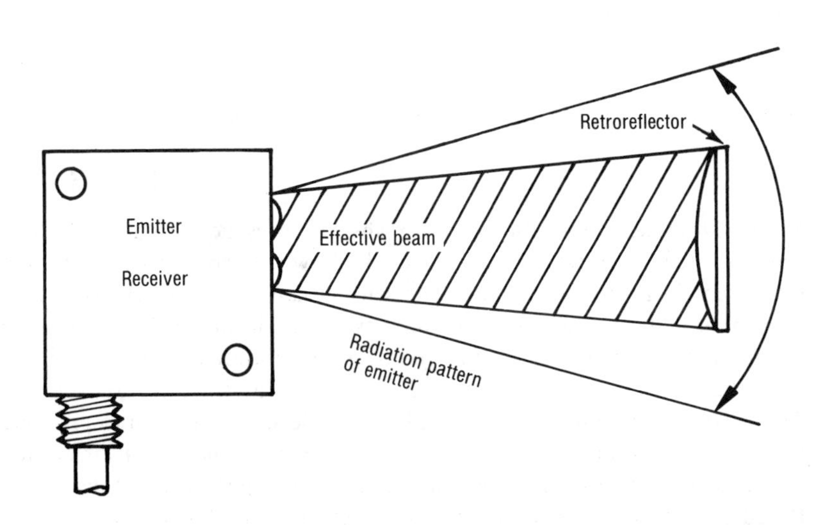

Figure 5-13
Effective Beam Diameter — Reflective Configuration
(courtesy Banner Engineering Corp.)

ELECTRO-OPTIC INTERFACE

The electro-optic interface is a transceiver that contains the light source, drive circuitry, photodetector, and amplification circuity. Most interfaces use a light-emitting diode (LED) as the light source. The interfaces are divided into two basic categories: modulated and nonmodulated.[5]

Nonmodulated units use a constant output level. They are capable of receiving high data rates as might be required in a high speed tachometer, but they are susceptible to false triggering due to ambient light being detected. Modulated systems pulse the LED at a relatively high frequency (typically several kilohertz). The detection system is designed only to receive signals at this pulse rate. Therefore, ambient light signals are eliminated in the detection scheme.

One major disadvantage of LED light sources is their inability to be used in sensing color. Light-emitting diodes transmit over a very narrow wavelength. IR light-emitting diodes, in particular, are unable to be used to distinguish color. Therefore, for sensing applications in which color discrimination is required, incandescent light sources must be used.

Figure 5-14 shows a typical modulated electro-optic interface.[4] The scanning mode is through-scan. The scanning block shows the oscillator, which pulses the LED light source, the receiver, and the amplification circuitry. The demodulator filters out unwanted light signals. The logic module drives the interface output.

APPLICATIONS

Most widespread applications are for absence/presence. Figure 5-15 illustrates a series of applications; (a) and (d) show the monitoring of product movement in a production line. The sensors can count objects as well as determine if there is a jam. Generally, as the environment becomes more severe, fiber optic sensors are favored over conventional photoelectric sensors. Figure 5-15(b) shows fiber optic sensors in a web detection scheme. The sensor can monitor for tears and other defects, in addition to precisely monitoring the web edge. Fiber optics have been used successfully in ejected part detection, vibrating feeder bowl control, and part-in mold detection. Because of the sensitive nature of such devices and the ability to detect small targets, fiber optic sensors have been used for thread detection as well as for other small objects (see Figure 5-15(c)). Targets as small as a few thousandths of an inch can be detected.

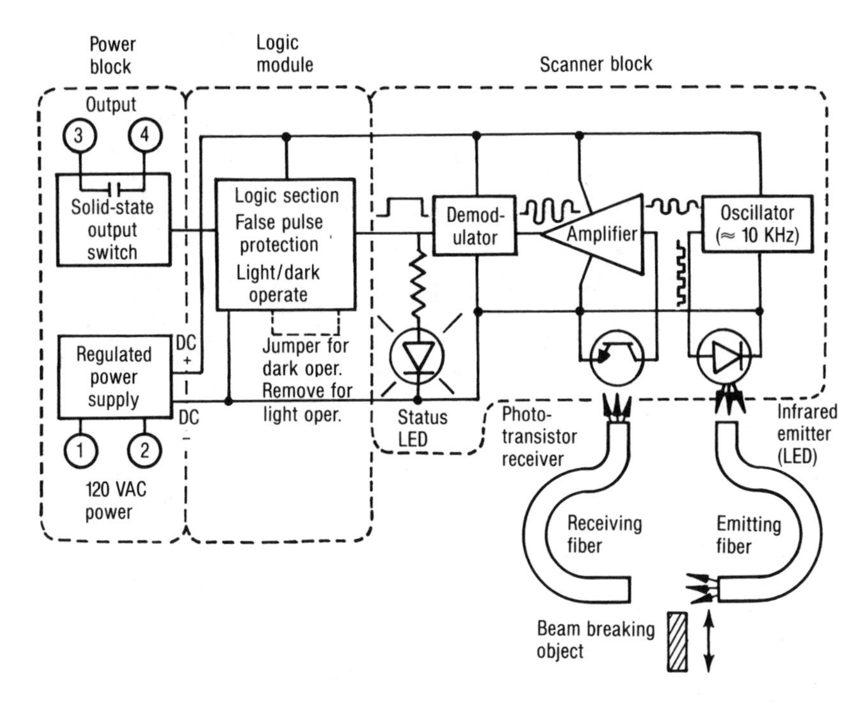

Figure 5-14
Typical Modulated Electro-optic Interface

(courtesy Banner Engineering Corp.)

Fiber optic absence/presence sensors have been used extensively as tachometers. A typical tachometer is shown in Figure 5-16. The device uses a bifurcated fiber optic probe with a reflective or absorption strip used as the target to provide a pulse train corresponding to rotation rate. In many applications where rotation rate determination is required, access is quite difficult. Fiber optics generally eliminate that problem.

Fiber optic tachometers have also been used in motor speed control. Figure 5-17 depicts a schematic representation of a speed control system. A specific application of a fiber optic tachometer is shown in Figure 5-18. The flywheel in the watt-hour meter is rotating 50 revolutions for the movement of one unit on the least significant dial indicator. The flywheel (rotating disk) has a black absorption strip for encoding purposes. The environment that the sensor must withstand is surprisingly severe, and this dictates the use of

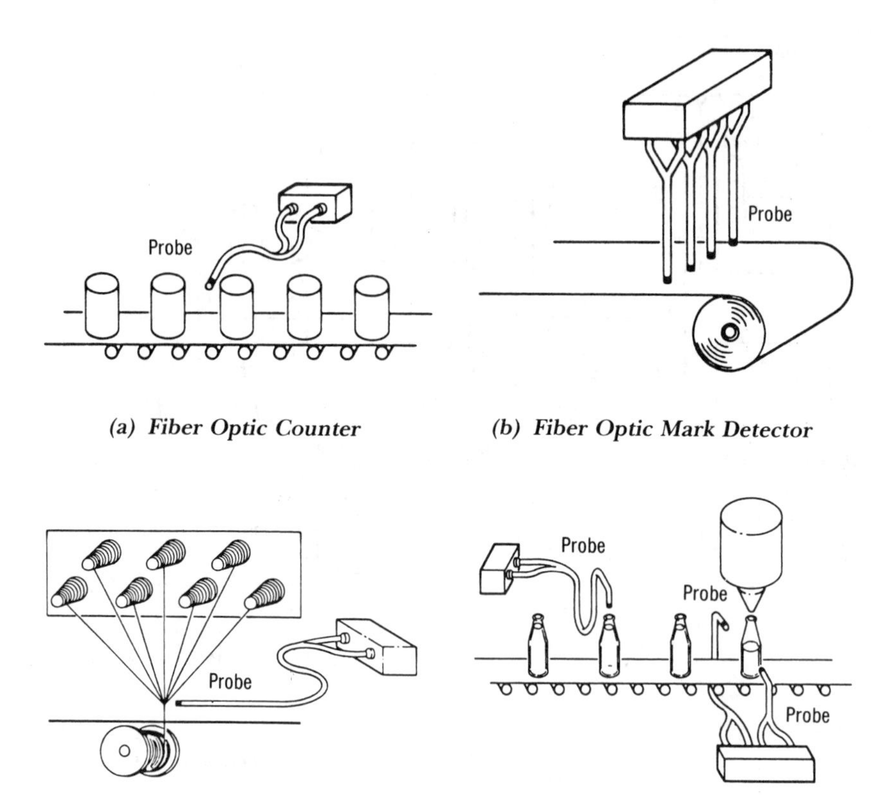

(a) *Fiber Optic Counter*

(b) *Fiber Optic Mark Detector*

(c) *Fiber Optic Thread Detector*

(d) *Fiber Optic Position Monitor*

Figure 5-15
Switch and Counter Applications

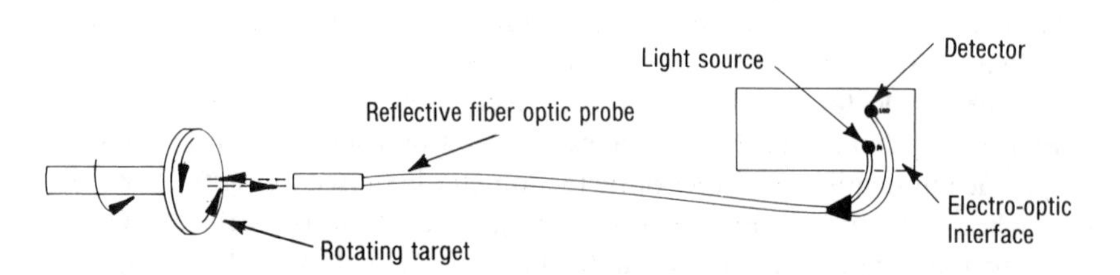

Figure 5-16
Fiber Optic Tachometer

fiber optics. The sensor/encoder can not be defeated by direct or indirect sunlight or by manual tampering. The relative humidity often approaches 100%, and the lensing effect of the glass housing in direct sunlight can result in temperatures exceeding 150 °C.

Another application that often requires fiber optic absence/presence sensors is hot object detection. In this case, if the object is over the incandescent

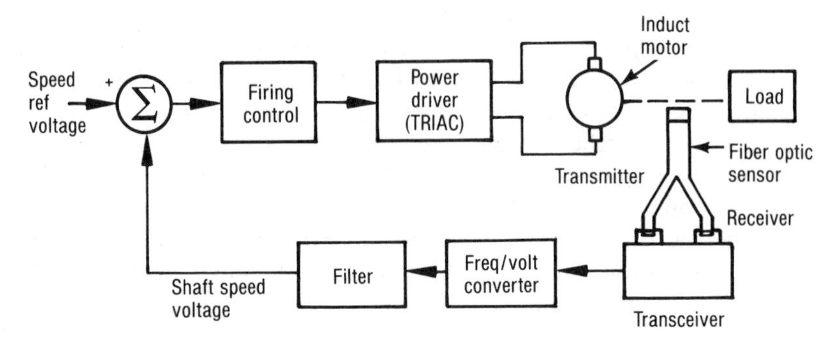

Figure 5-17
Fiber Optic Speed Control System

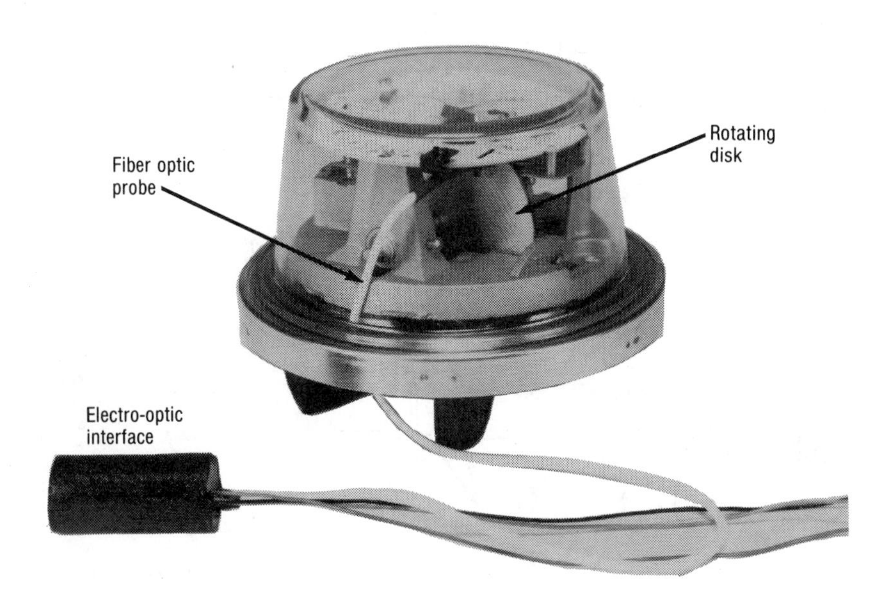

Figure 5-18
Fiber Optic Tachometer Used in a Watt-Hour Meter

temperature, the object itself provides the light source. Therefore, only the receive probe in a through-scan configuration is needed. Unfortunately, most conventional bundle or single-fiber probes are limited to about 200 °C because polymeric protective coatings cannot withstand higher temperatures. Special adapters, such as quartz rod extenders, are often required; these act like a large fiber optic to handle the heat. The stiff large quartz rods do not require protective coatings.

As mentioned previously, there is a large potential switch application in the machine tool area. Sensors can be used to gauge parts or determine if the cutting tool is present. In almost all situations there is a high volume of liquid coolant or airborne contamination flowing in the vicinity of the cutting tool. The coolant and the debris can foul the optics by essentially blinding the optical probe.[7,8]

Commercial systems have addressed this problem. The probes can be kept clean even in the presence of high volumes of coolant by using an air wipe concept as shown in Figure 5-19. With a slight positive pressure over

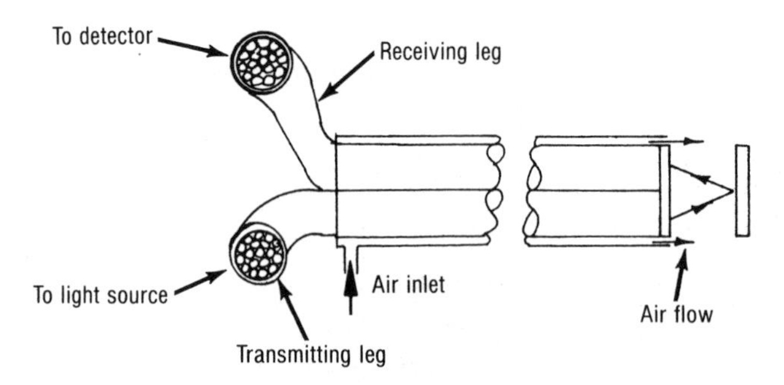

(a) Reflective System

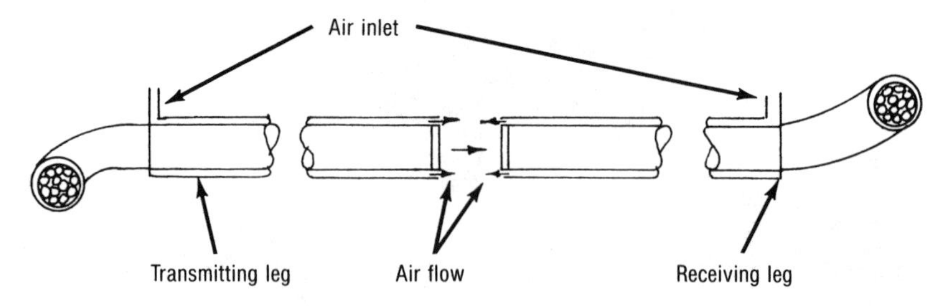

(b) Transmissive System

Figure 5-19
Fiber Optic Sensors with Air Wipes

the end of the probe, coolant and debris are kept off the optical surface as shown in Figure 5-20. The system is further enhanced if a spoiler causes a turbulent swirling air flow, which enhances the cleaning action. In normal field applications, most of the machinery that uses cutting tools has compressed air sources in close proximity so that installation is simple.

Widespread use of digital fiber optic switches and counters has been limited due to functionality. Most applications have been limited to absence/presence, since these are the simplest sensors to develop and use. Digital sensing is now expanding to liquid level, temperature, and pressure[7] (see Figure 5-21).

For the expanded digital sensor, the target provides the sensing function. If a target can be fabricated such that its position or state relative to the probe end is a function of some physical parameter to be sensed, then physical property sensors can be fabricated. All the sensors shown in Figure 5-21 are using a reflective technique. Consider a reflective bimetal ele-

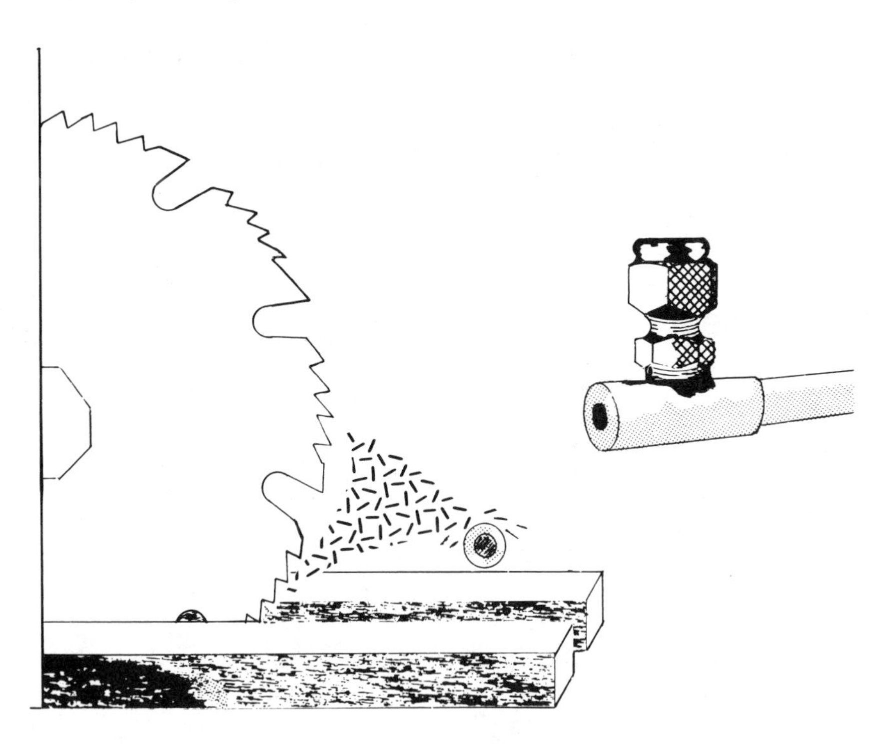

Figure 5-20
Air Wipe for Dirty Environments

(reprinted by permission from EOTec Corporation)

ment as one such target. As the surrounding area reaches its set point temperature, the element moves abruptly. The distance between the probe and target shown in Figure 5-22 changes correspondingly, affecting the amount of light reflected back to the probe, which provides the basis for switching. Pressure sensing via fiber optics is provided by a method analogous to the bimetal temperature element. A flexible pressure-sensitive diaphragm with a reflective inner surface varies its distance from the fiber optic probe tip in response to a pressure input, as shown in Figure 5-23. Using a snap diaphragm, a pressure set point is achieved.

Another technique for utilization of fiber optic probes involves the use of a prism tip for liquid level sensing[1] (see Figure 5-24). Light traveling down one leg of the probe is totally internally reflected at the prism air interface. Note that air (index of refraction = 1) acts as a cladding material around the prism. As the prism contacts the surface of a liquid, light is stripped from the prism, resulting in a loss of energy at the detector. With the proper electronic circuitry, discrimination can be achieved between liquid types, such as gasoline and water, by the amount of light lost from the system, a function of the index of refraction of the liquid.

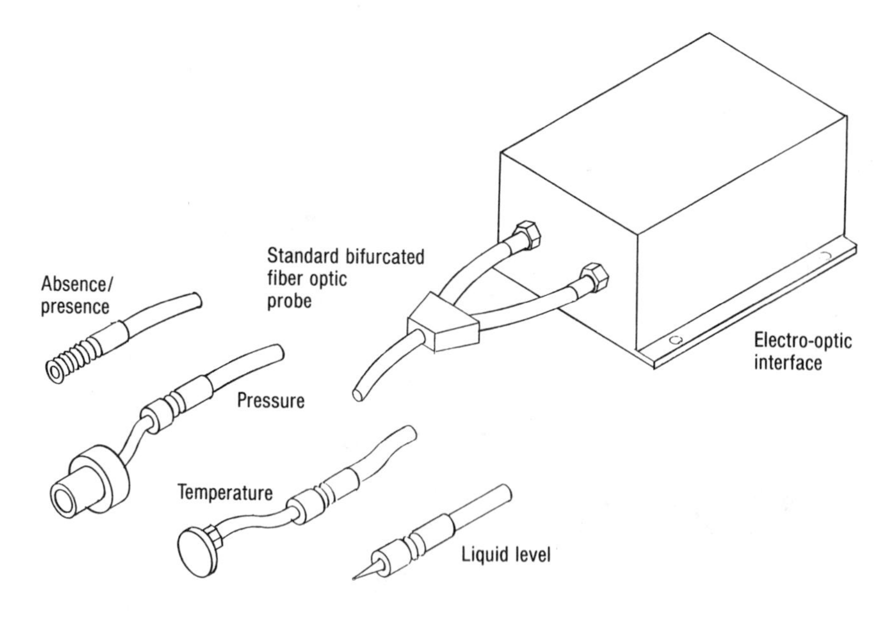

Figure 5-21
Expanded Sensor Functionality

With the exception of the liquid level switch, the devices used for physical property sensors, which were discussed in the previous examples, were opto-mechanical in nature. Intrinsic fiber optic physical property switches are also possible. These devices do not require the use of a mechanical transducer. A multitude of materials will change their optical properties with

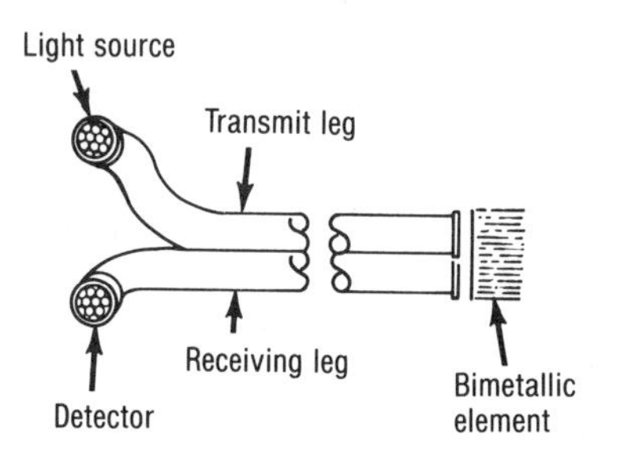

(a) Concept

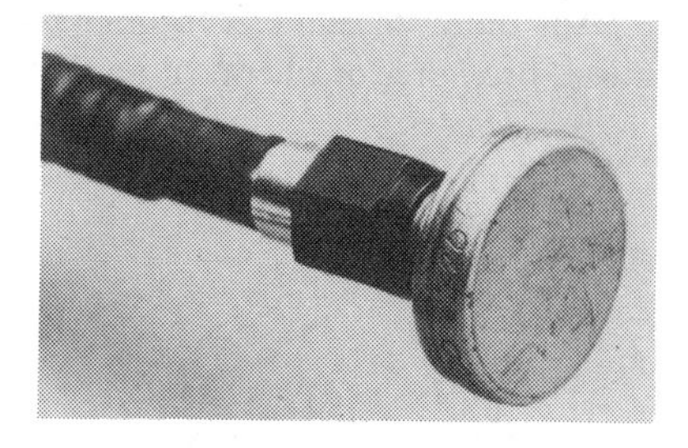

(b) Actual Device (EOTec)

Figure 5-22
Fiber Optic Temperature Switch
(reprinted by permission from SENSORS, ©1984 Helmers Publishing, Inc. and from EOTec)

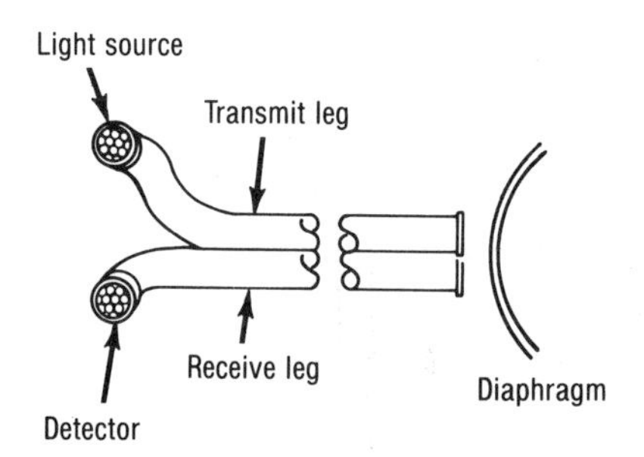

(a) Concept

(b) Actual Device (EOTec)

Figure 5-23
Fiber Optic Pressure Switch
(reprinted by permission from SENSORS, ©1984 Helmers Publishing, Inc. and from EOTec)

environmental changes and can interact with a transmissive or reflective fiber optic system. Currently, most devices are experimental, but a few concepts are worth mentioning: polymeric materials can abruptly change transmission and/or reflectivity at rather precise temperatures and can function as the basis for a temperature switch; liquid crystals are sensitive to temperature and pressure with abrubt transmission changes and potentially can function as temperature or pressure switches.

Any material that undergoes a physical change of state at a given temperature, pressure, or chemical level is a candidate as a switch component.

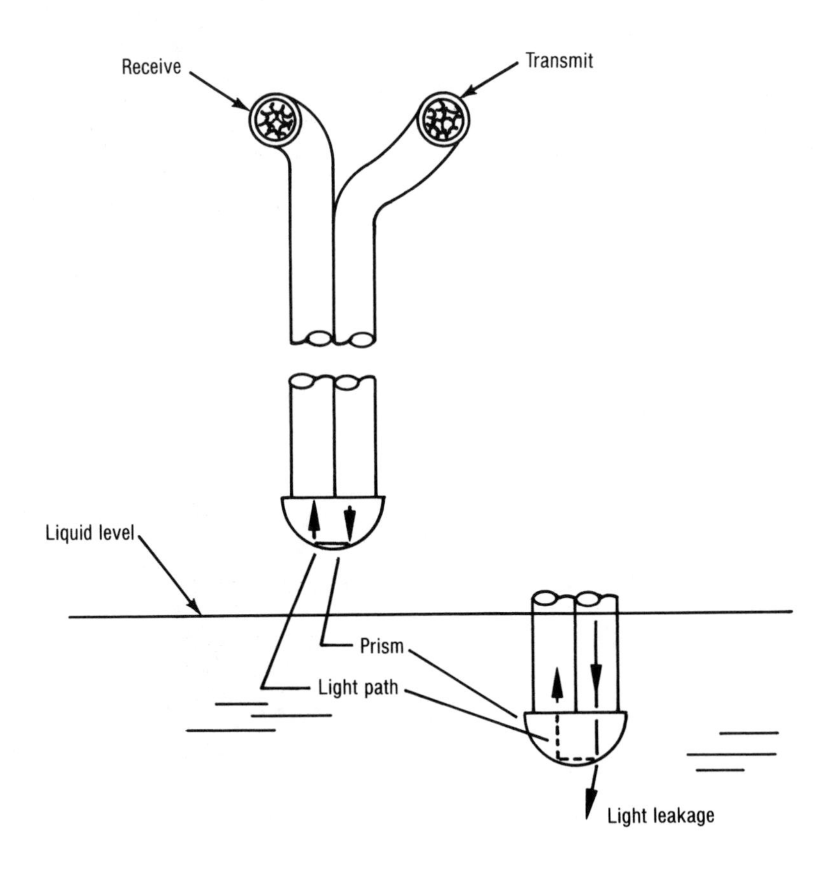

Figure 5-24
Fiber Optic Liquid Level Switch

REFERENCES

1. "MicroSwitch," 1984, MP Series Catalog.

2. Banner, 1983, *Photoelectric Control and Reference Manual.*

3. Damuck, Jr., W. E., 1984, "A Comparison of Photoelectric and Fiber Optic Sensors," *Proceedings of the ISA International Conference — Houston, TX*, Vol. 39, pp 303-12.

4. Veeder Root, 1983, *Photoelectric Sensors — Specification Requirements.*

5. Fayfield, R. W., 1982, "Fiber Optics and Photoelectric Sensing, A Good Combination," *Instrumentation and Control*, pp 45-49.

6. Krohn, D. A., 1983, "Fiber Optic Sensors in Industrial Applications — An Update," *Proceedings of the ISA International Conference — Houston, TX*, Vol. 38, pp 877-90.

7. Krohn, D. A., 1986, "Fiber Optic Sensors," *Solid State Sensors* Workshop Sponsored by IEEE.

8. Krohn, D. A., 1985, "Field Experience with Fiber Optic Sensors," *Proceedings of the ISA International Conference — Philadelphia, PA*, Vol. 40, pp 1051-61.

9. EOTec, 1986, *Fiber Optic Sensors* — A Complete Reference Manual.

10. Coulombe, R. F., 1984, "Fiber Optic Sensors Catching Up with the 1980's," *Sensors*, pp 5-8.

Displacement Sensors

6

INTRODUCTION

Fiber optic displacement sensors will play an increasingly larger role in a broad range of industrial, military, and medical applications. The driving force will be the many advantages of fiber optics as listed in previous chapters. Two particular advantages include the potential for extremely accurate non-contact sensing and the possibility of incorporating the optical sensors permanently in composite structures.

The basic fiber optic intensity-modulated sensing concepts outlined in Chapter 3 as well as interferometric sensors can measure displacement. Two concepts, however, have provided the primary approach to displacement sensors: reflective and microbending sensors. This chapter will focus on a more detailed review of how these sensors function in a displacement sensor mode and how they can be applied to various applications.

REFLECTIVE TECHNOLOGY

The basic concept of a reflective sensor is shown in Figure 6-1. The sensor is comprised of two fiber optic legs (bundle or single fiber). One leg transmits light to a reflecting target; the other leg traps reflected light and transmits it to a detector. The intensity of the detected light depends upon how far the reflecting target is from the fiber optic probe. The basic response curve is shown in Figure 6-2.[1-3] The curve shows a maximum with a steep linear front slope and a back slope that follows a $1/R^2$ dependence, where R is the distance between the tip of the fiber optic probe and the reflecting surface.

The curve is easily understood if the geometric optics are considered. Light exits the transmitting fibers in a solid cone defined by the numerical

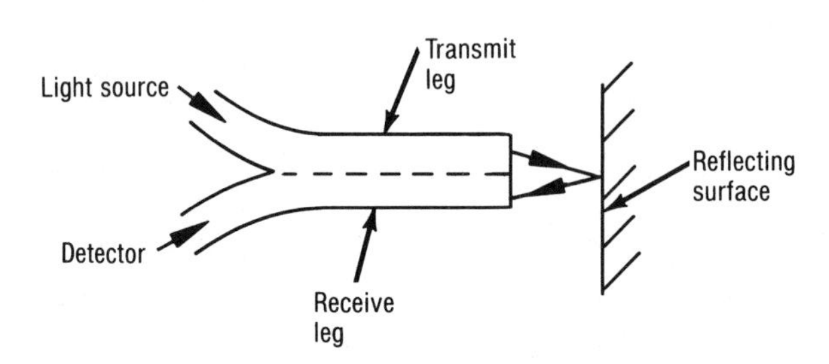

Figure 6-1
Reflective Fiber Optic Sensors

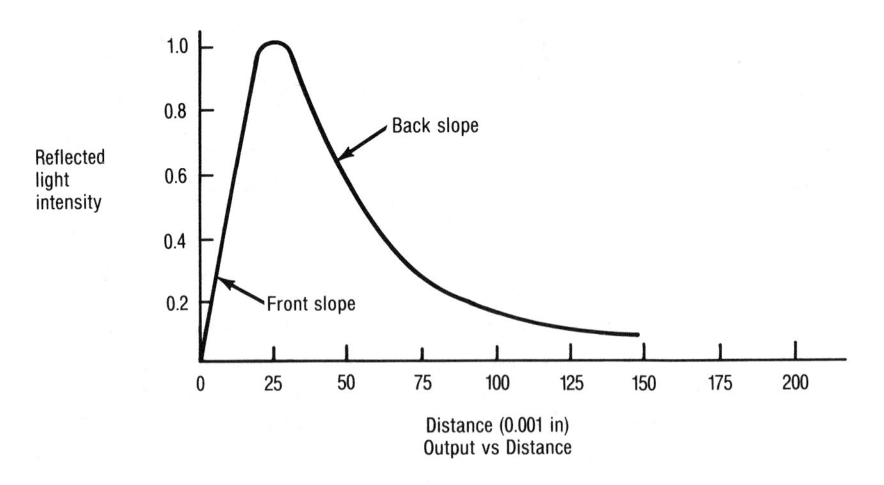

Figure 6-2
Reflective Fiber Optic Sensor Response Curve

aperture. The spot size hitting the target is given by (assuming the fiber is small in relation to R):

$$2R \tan \theta = D \tag{6-1}$$

where θ is the half angle between the normal to the fiber exit surface and the exit divergence cone (numerical aperture); and D is the spot diameter.

Since the angle of reflection is equal to the angle of incidence, the spot size that impinges back on the fiber optic probe after reflection is twice the size of the spot that hits the target initially. As the distance from the reflecting surface increases, the area of the spot increases in a manner directly proportional to R^2 (see Figure 6-3). The amount of detected light is inversely proportional to the spot area or $1/R^2$, since the receiving fiber is fixed in size and less of the fiber face intersects the returning light as it expands with distance. As the probe tip comes closer to the reflecting target, there is a position in which the reflected light rays are not coupled into a receiving fiber. At the onset of this occurrence (position 2), a maximum forms, which drops to zero as the reflecting surface contacts the probe, as shown in Figure 6-4.

It is clear from an explanation of how the curves are generated that the output can be varied by using various optical configurations, as shown in Figure 6-5. As a result, the dynamic range and sensitivity can be tailored for

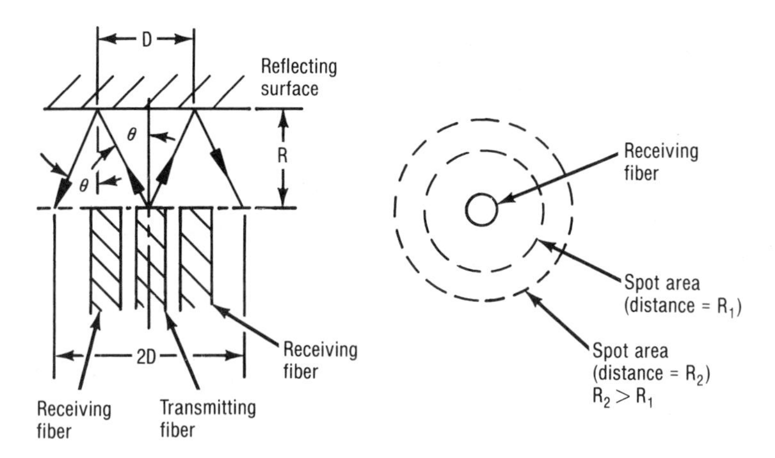

(a) Fiber Target Geometry *(b) Reflected Spot Area*

Figure 6-3
Back Slope Considerations

a specific application. As an example, the random configuration has, on the average, the closest transmit fiber to receive fiber spacing. Therefore, at close target distances it would have the highest sensitivity, and the lowest sensitivity at long distances. The fiber pair probe has the largest transmit fiber to receive fiber spacing (center-to-center) and, therefore, is the least sensitive at close target distances. It should be noted that in a single-fiber configuration the front slope disappears because, when the reflecting surface is approached, light continues to be reflected into the same fiber, which is functioning in both transmit and receive modes. This is true even to the point of contact. The front slope also disappears if a fiber optic lens (Selfoc)* is added to the probe, since the lens prevents the probe tip from coming close enough to generate the front slope. The output of a fiber pair probe with a Selfoc lens is shown in Figure 6-6. A Selfoc lens is a fiber optic self-focusing lens.

Dual probes provide a means for increased sensitivity. Consider the arrangement in Figure 6-7.[5] The probe distances are set so that the reflected intensity is the same for probe A in the front slope and probe B in the back

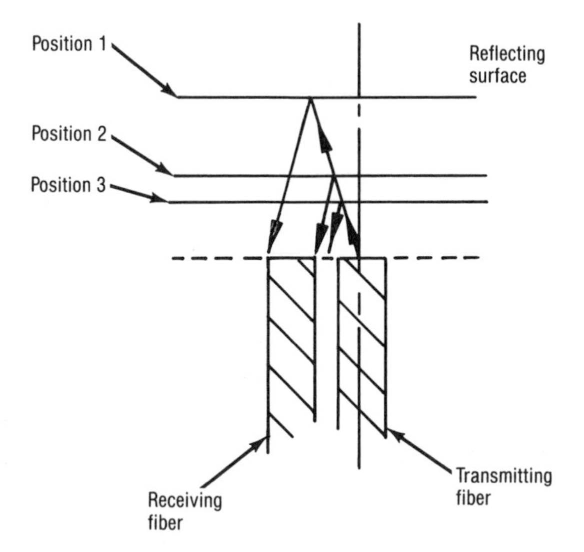

Figure 6-4
Front Slope Considerations

*Trademark, Nippon Sheet Glass.

slope. As the target moves closer, the detected intensity for probe A decreases while the detected intensity for probe B increases.

The difference between the two readings provides a larger output than either probe used singly and, therefore, increased sensitivity. In addition, displacement direction information is obtained in addition to magnitude. A similar effect can be obtained if the probes are equidistant from the target but on opposite sides, as shown in Figure 6-8(a). Depending on whether the probes are both set on the front or the back slope, if the target moves in

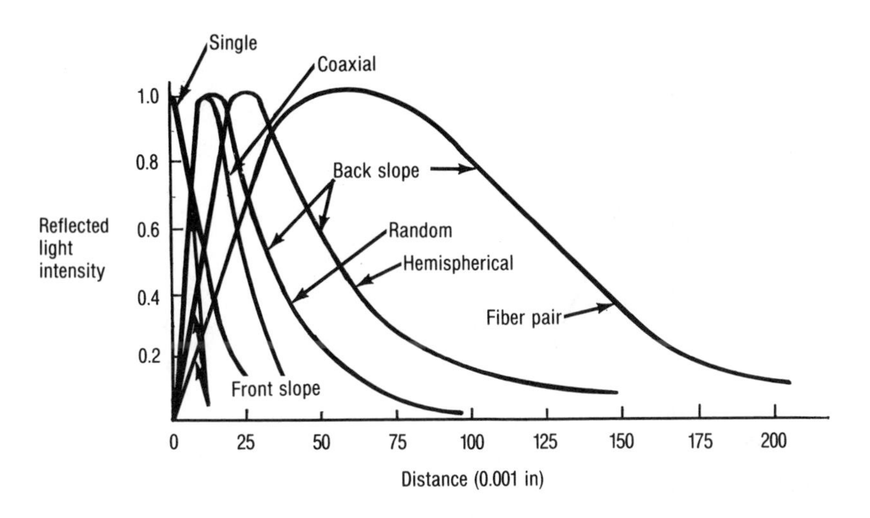

(a) Output vs Distance

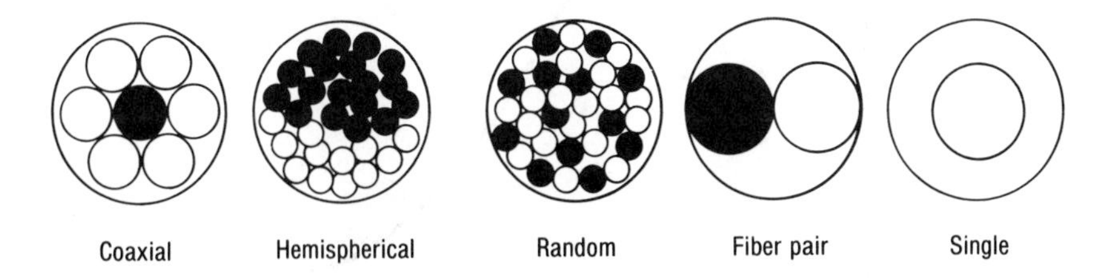

(b) Probe Configuration

Figure 6-5
Reflective Fiber Optic Sensor Response Curve for Various Configurations

either direction, the measured displacement of one probe will increase while the other decreases. Figure 6-8(b) shows the differential output (output of probe A minus output of probe B) for the probes positioned on the front slope.

The major disadvantage of the sensors described is the limited dynamic range. While sensitivities approaching 1 microinch are possible, the dynamic range is limited to about 0.2 inch. Many applications require that the sensor be used at a much greater distance (up to several inches) from the reflecting target. Using a lens system in conjunction with a fiber optic probe, the dynamic range can be expanded to 5 or more inches,[1,2] although the

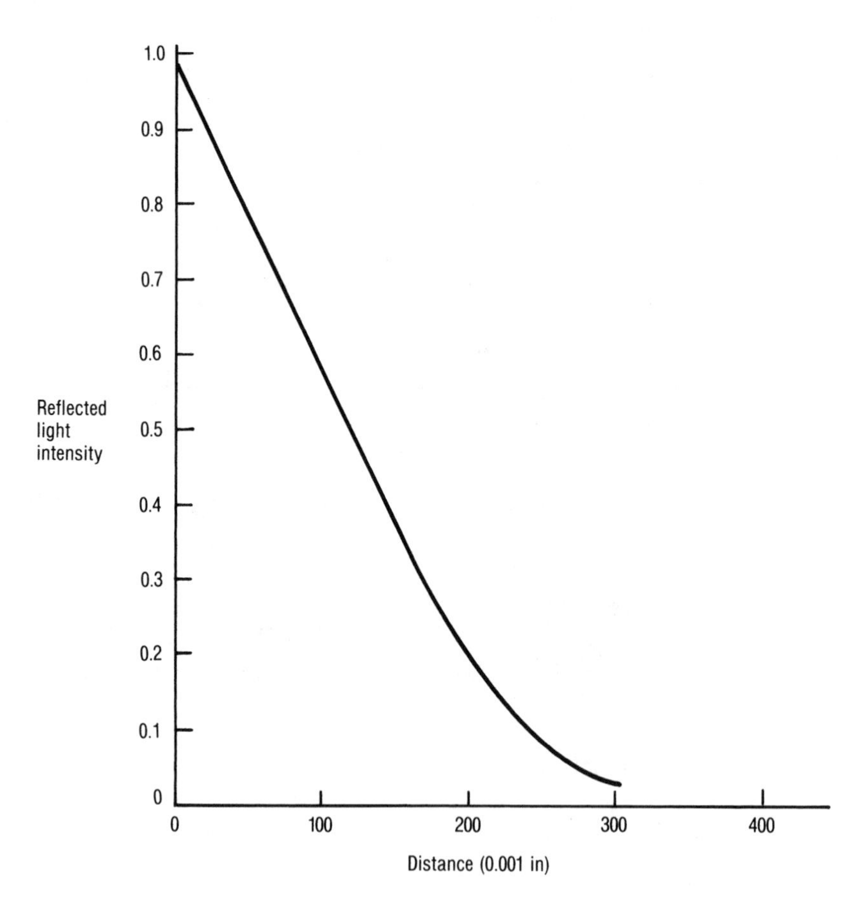

Figure 6-6
Response Curve for Fiber Pair with Selfoc Lens

sensitivity will be correspondingly reduced. The response curve for various focal length settings is given in Figure 6-9 using a random probe. With the exception of the curve for an infinite focal length, the curve now has two maxima. The explanation for the trough at the focal length is straightforward. The object is in exact focus at that point; light from a particular fiber

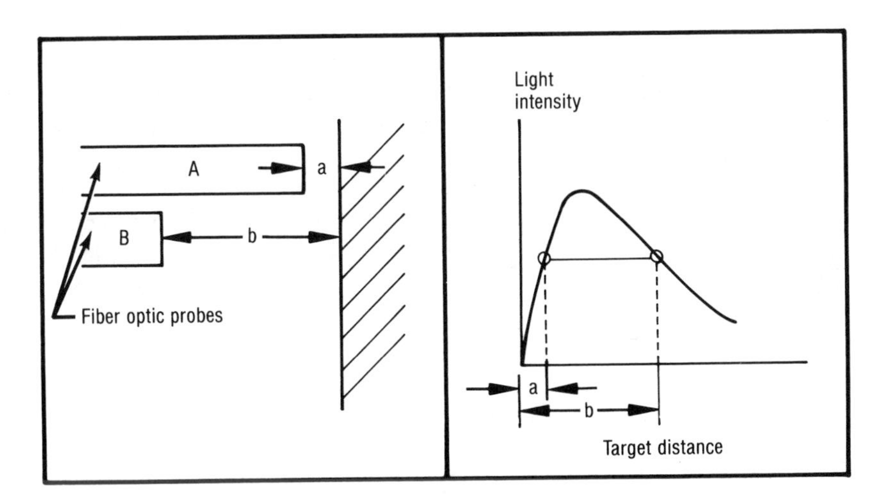

Figure 6-7
Dual Reflective Probes Same Side of the Target

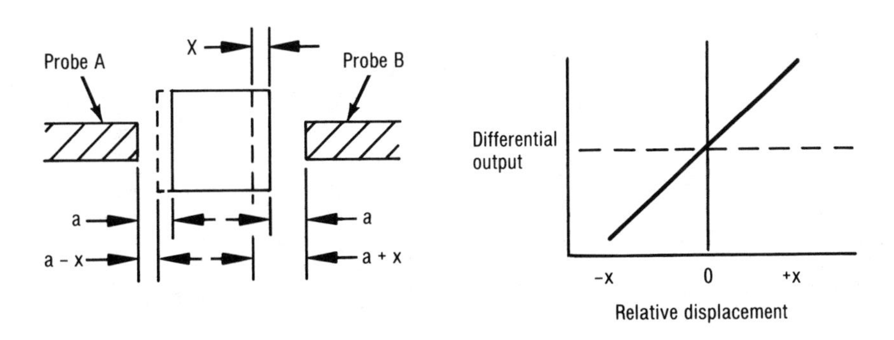

(a) Probes on Opposite Sides

(b) Probes on Front Slope

Figure 6-8
Dual Probes on Opposite Sides of the Target

has a very small spot size and reflects back on itself; therefore, the detected intensity is minimized. As the object moves in and out of focus, the defocused spot couples more reflected energy into an adjacent receiving fiber, increasing the level of detected light. The highest sensitivity is on the front slope of the trough. The sensitivity and dynamic range can be adjusted by changing the focal length of the lens. In essence, it is possible to be 3 inches from the surface and detect movement to an accuracy of 0.001 inch over a dynamic range of 1 inch using a 3 inch focal length lens.

There are some difficulties with reflective fiber optic displacement sensors. The sensor integrates distance data with changes in reflectivity of the target and angularity of the target. In an effort to characterize the behavior, a micro-finish comparator was used in which lapped surfaces were prepared by cutting tools of varying coarseness. A $2L$ surface is nearly specular. The diffuse nature of the surfaces increases from $4L$ to $8L$ to $16L$, with $32L$ being the most diffuse. A plot of reflected light intensity versus distance for

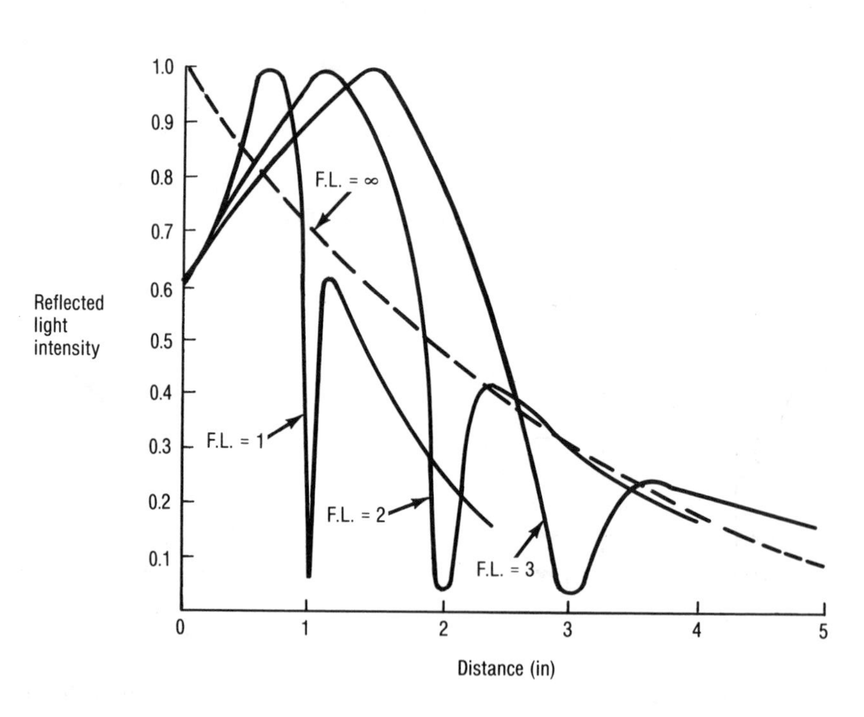

Figure 6-9
Response Curve for Reflective Sensor Coupled with Lens System for Various Focal Lengths

different surfaces is given in Figure 6-10. As the surface becomes rougher, the reflected light intensity drops substantially. However, the general shape of the curve remains constant with the maximum fixed at a given distance. If the same target is always being used in an application and the surface remains clean, the reflectivity variations are eliminated. On the other hand, if different surfaces are being measured, a reference probe will be required to compensate for the reflectivity variations.

The sensor is also sensitive to rotation of the reflecting target, as shown in Figure 6-11. The errors appear to be small ($\pm 3\%$) for rotations of $\pm 5°$ about the normal. Decreases in light intensity become quite pronounced at large angles.

A major advantage of a reflective fiber optic sensor is that contact is not required for measurement. There are applications, however, in which surface reflectivity can change due to contamination, so that a closed system is required. Using an optical version of a linear voltage displacement transformer (LVDT), contact of the part to be measured is required by the mechanical plunger but all other problems are eliminated. The optical sensing occurs in a closed environment in which the sensor, through reflections at the mirror, tracks the movement of the plunger.

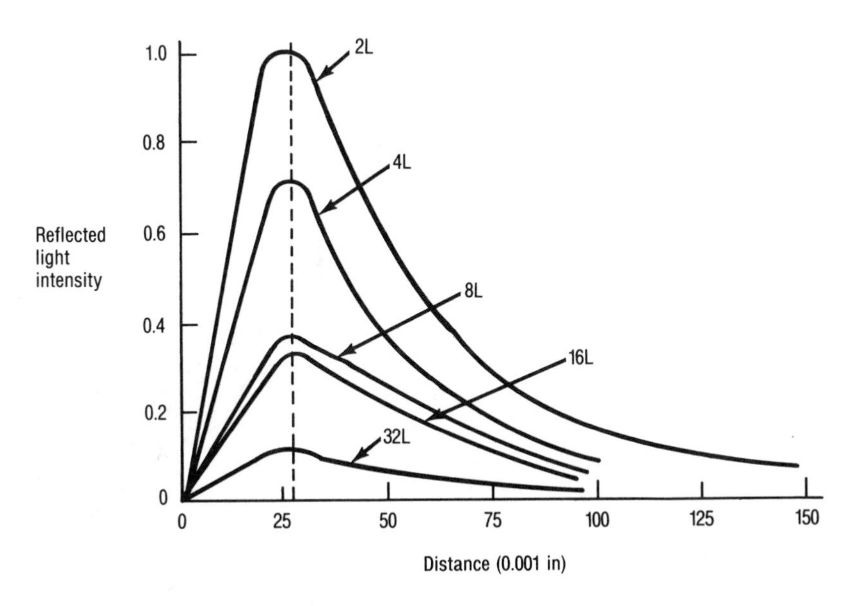

Figure 6-10
Effect of Target Reflectivity Changes

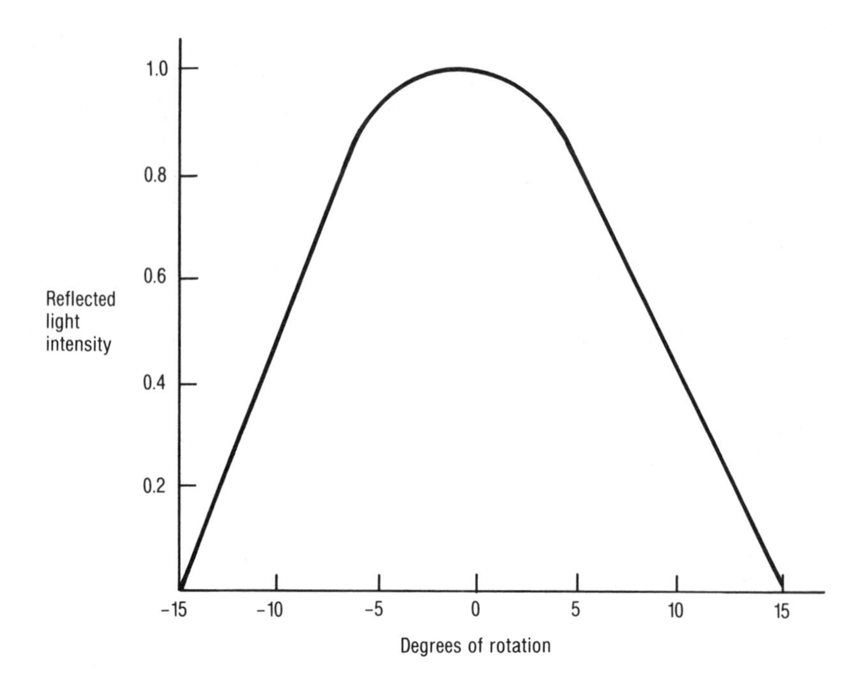

Figure 6-11
Effect of Target Angularity

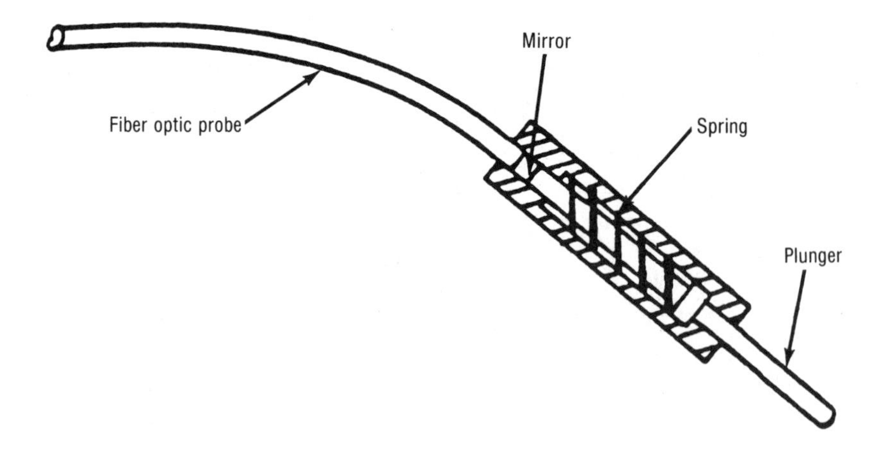

Figure 6-12
Optical Version of an LVDT

MICROBENDING TECHNOLOGY

All optical fibers will radiate energy when bent.[6] The energy distribution in a bent fiber is shown in Figure 6-13. Theoretically, the energy field in the cladding extends to infinity. Therefore, at some radiation distance, X_r, the energy is implied to propagate at a velocity greater than the speed of light due to the longer travel path. Since this is not possible, the waveguide effect ceases. Subsequently, the energy is lost to radiation.

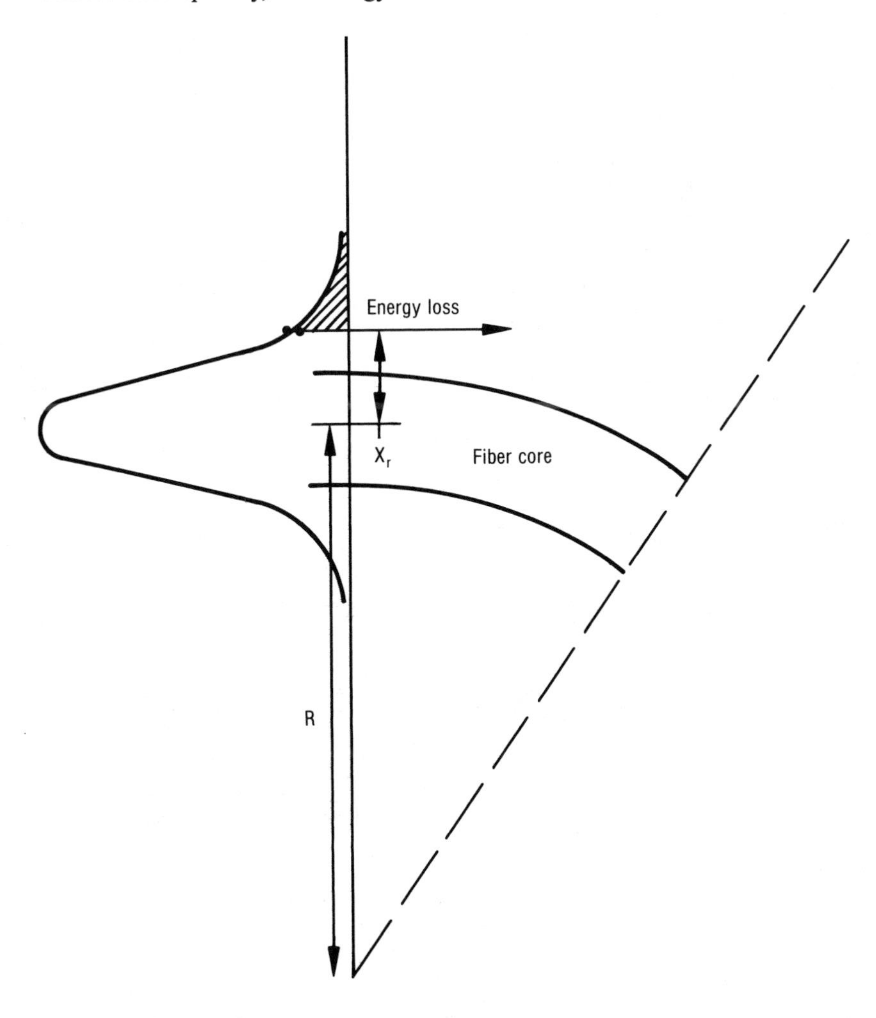

Figure 6-13
Energy Distribution in a Bent Fiber

The effect of microbending is shown in Figure 6-14.[4] As the perturbing bend points contact the fiber, energy is lost.

The parameters that influence microbending loss include fiber numerical aperture, core size, core-to-cladding ratio and periodicity of fiber deformation. When a periodic microbend is introduced, mode coupling occurs between modes with longitudinal propagation constants β and β^1.[7-9] Therefore:

$$\beta - \beta^1 = \frac{2\pi}{\Lambda} \tag{6-2}$$

where Λ is the wavelength of periodic bending. The difference in propagation constant for adjacent modes is given by:

$$\beta_{m+1} - \beta_m = \frac{2.83 \text{ N.A. } (m)}{a\, n_1 (M)} \tag{6-3}$$

where N.A. is the numerical aperture, n_1 is the core index, a is the core radius, m is the modal group label, and M is the number of modal groups. Equating Equations 6-2 and 6-3 and solving for Λ gives:

$$\Lambda = \frac{a\, n_1 (M)}{1.414 \text{ N.A. } (m)} \tag{6-4}$$

Higher-order modes (M) are coupled with small Λ and lower-order modes (m) are coupled with large Λ.

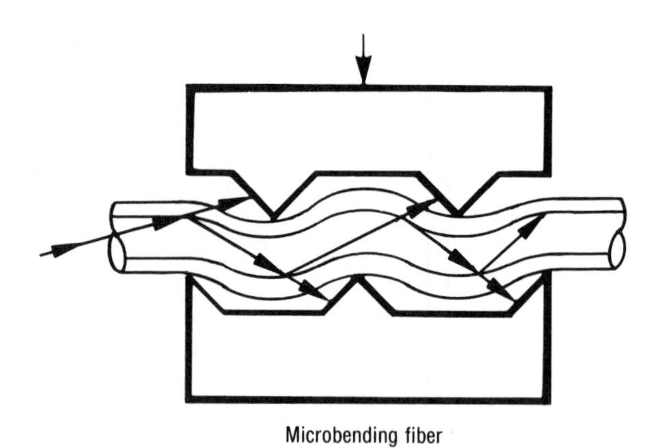

Microbending fiber

Figure 6-14
Light Leakage as a Result of Microbending

Microbending transducers are based on coupling core modes to radiated modes. The greatest sensitivity occurs when the highest-order core modes are made to radiate. This situation can be mathematically approximated when $M = m$. Therefore, Equation 6-4 is reduced to:

$$\Lambda = \frac{a\, n_1}{1.44\ (\text{N.A.})} \tag{6-5}$$

As the numerical aperture increases, the periodic perturbation spacing must decrease for maximum sensitivity. As core size deceases, the spacing must also decrease. Generally, high N.A. fibers or small core fibers guide light more strongly and require more severe bending to lose light. For a given fiber diameter, increases in the core-to-cladding ratio increase microbending sensitivity since, for a given displacement, the large core will have more bending and, therefore, more light leakage.

It has been shown that light loss is strongly dependent upon the microbending periodic perturbation, Λ. As this period approaches the magnitude of the ray period of light propagating in the fiber, a strong resonance is possible with the light loss showing a maximum.(10) Figure 6-15 plots the change in transmission relative to the change in displacement, $\Delta T/\Delta X$, versus Λ. The data was obtained with a step index fiber under conditions of constant load using the fixture in Figure 6-16. The primary peak corresponds to the periodicity that couples the highest-order modes to radiant modes. As the period increases, the sensitivity decreases corresponding to the difficulty associated with coupling low-order modes.

Microbending sensors can be divided into two categories:(12) bright field and dark field. Bright field sensors measure the light level transmitted in the fiber, the change being associated with microbending losses. A typical bright field configuration is shown in Figure 6-16. Dark field sensors measure the light lost through the fiber cladding. Dark field configurations are shown in Figures 6-17 and 6-18.

Using the system described in Figure 6-16, light was injected into the fibers at different angles. The microbending fixture was under constant load; therefore, the fiber microbending is constant. It is clear that the higher incident angle light, which corresponds to higher-order modes, resulted in greater microbending loss, as shown in Figure 6-19.(13)

One major potential application for microbending sensors is displacement measurement. Figure 6-20 shows transmission versus displacement using the bright field mode. As discussed in Chapter 3, the response curve has three distinct regions. In the first region, the compliant coating absorbs the initial displacement movement and the fiber bending is partially limited. The small bending level results in only the leakiest modes radiating. The second region gives a linear response over about 60% of the transmission range. This region is normally used for sensing. As the displacement increases further, light depletion in the higher-order modes occurs and sensor sensitivity is greatly reduced.

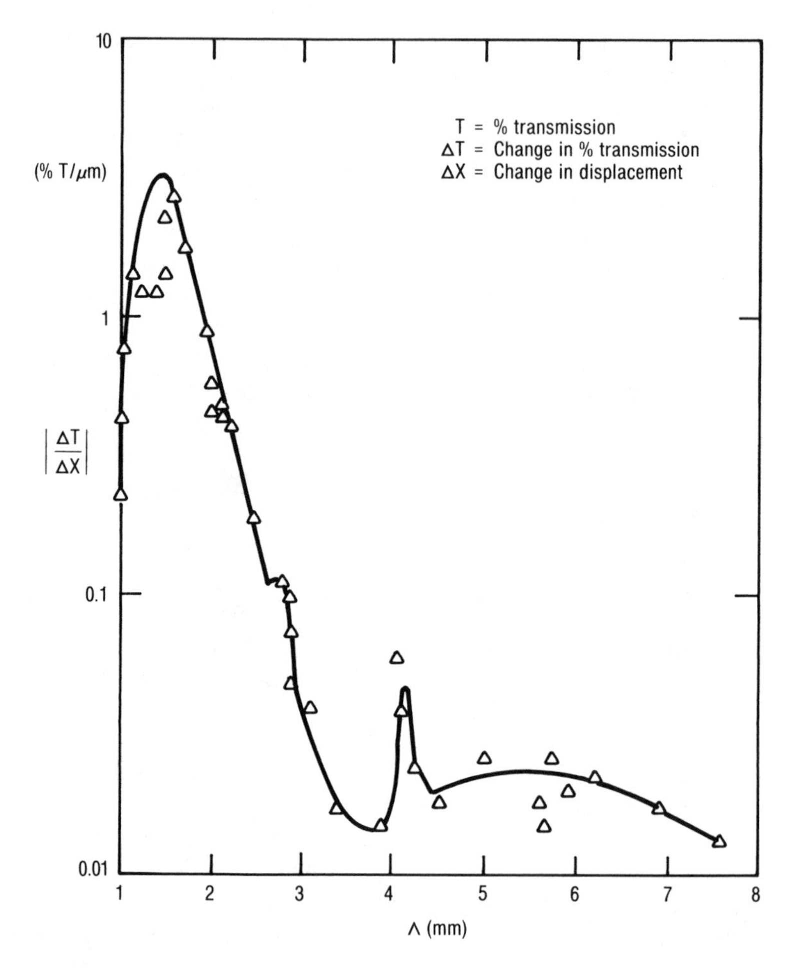

Figure 6-15
Displacement Sensitivity versus Λ

(© 1982 IEEE; reprinted with permission)

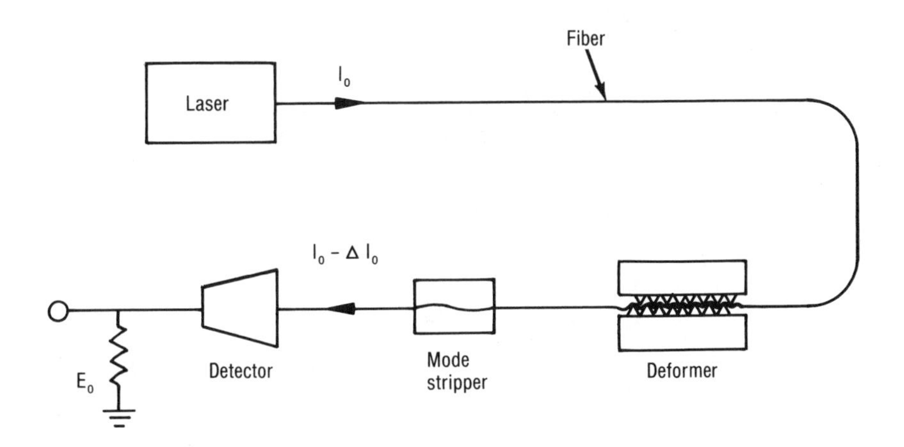

Figure 6-16
Typical Bright Field Microbending Sensor Configuration
(reprinted by permission from Dynamic Systems, Inc.)

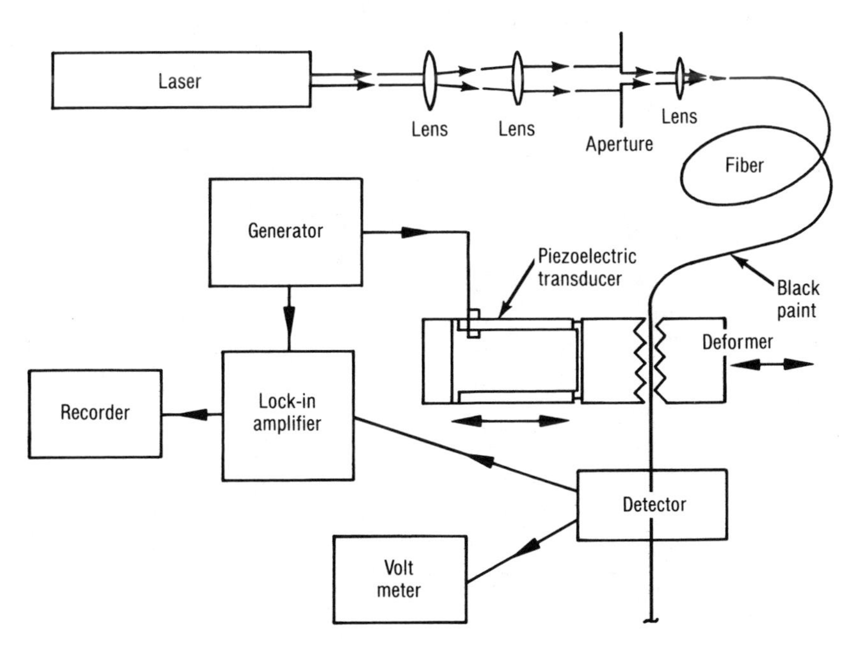

Figure 6-17
Experimental Set-up for Studying Displacement Microbending Sensor
(reprinted by permission from the Optical Society of America)

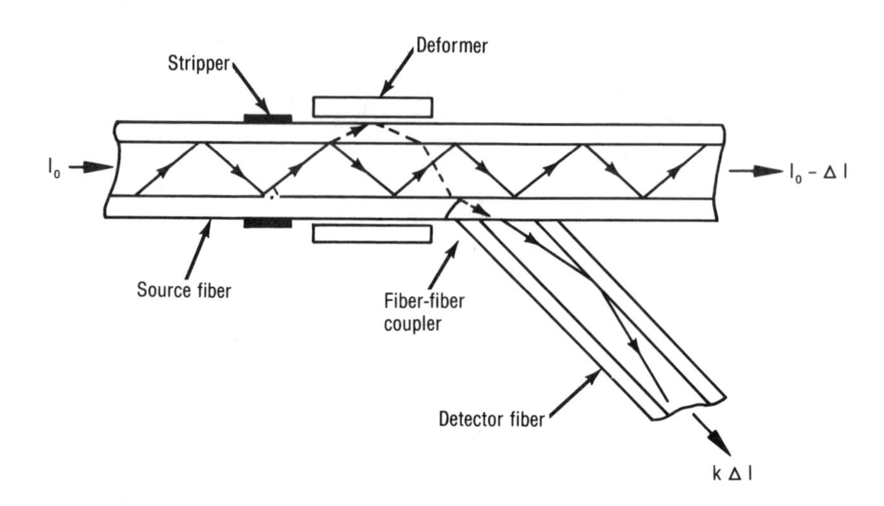

Figure 6-18
A Darkfield Microbend Intensity-type Fiber Optic Sensor
(reprinted by permission from Dynamic Systems, Inc.)

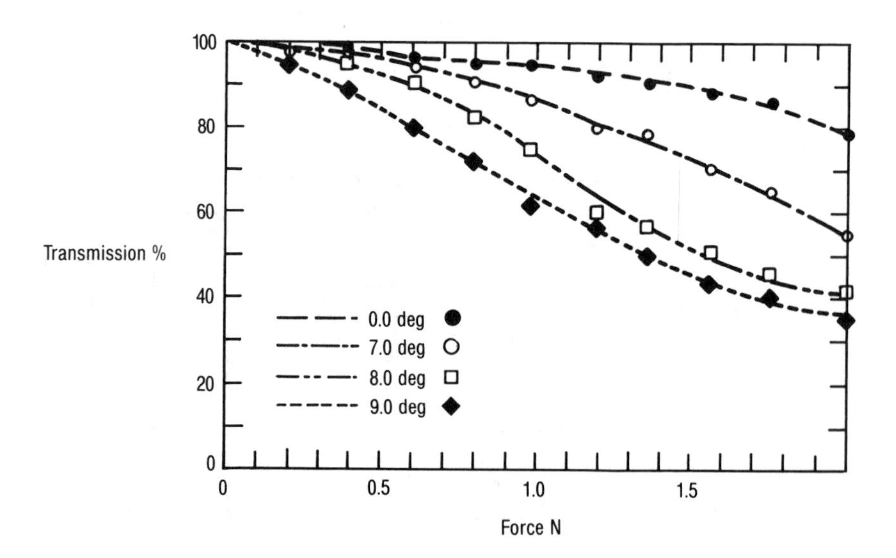

Figure 6-19
Microbending Loss versus Angle of Incident Light

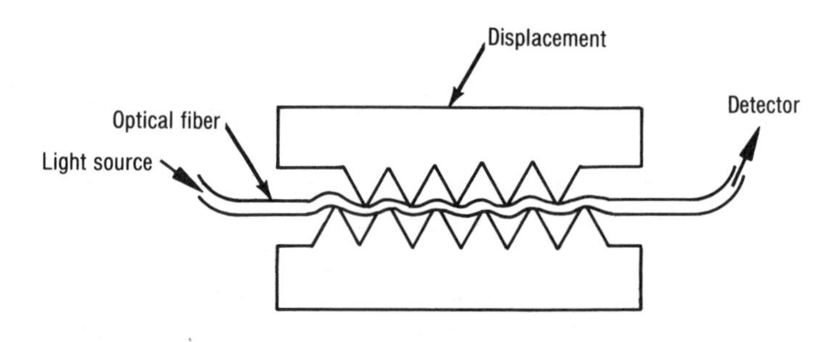

(a) *Sensor Arrangement*

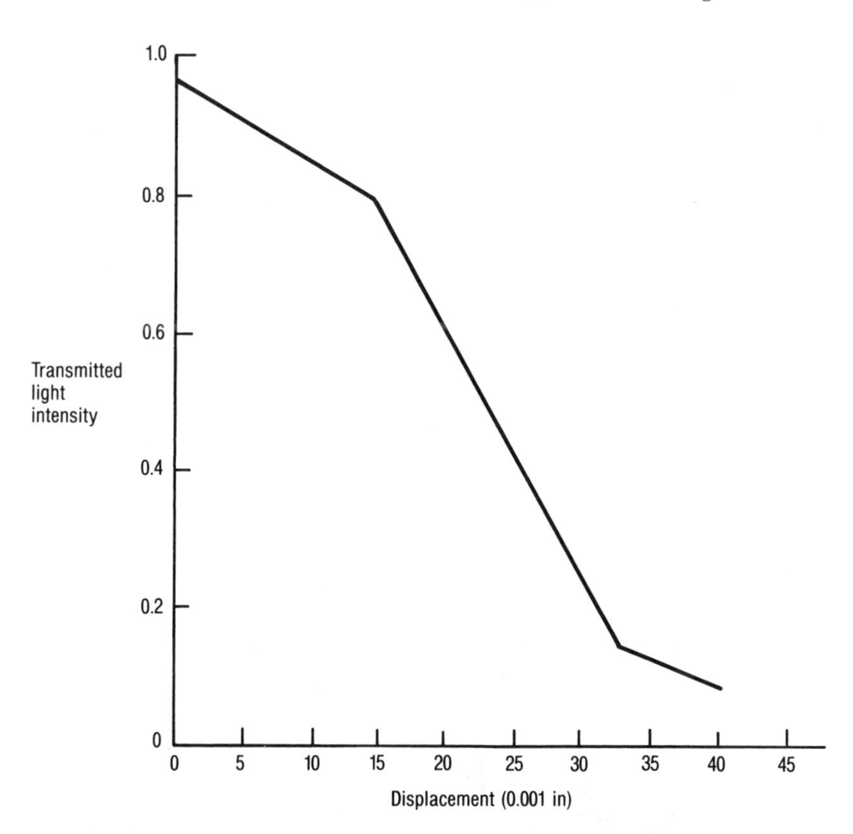

(b) *Output versus Displacement*

Figure 6-20
Transmission versus Displacement (Bright Field Microbending)

It is anticipated that in the dark field mode the collection efficiency is greatly reduced. Therefore, the response curve is similar in shape, but the mirror image is reduced by some light collection efficiency scale factor, as shown in Figure 6-21.

A major consideration in microbending sensors is stability. Most fibers are coated with a polymeric material as a protective mechanical buffer. It has been observed that such coatings can flow under heavy loads and/or increased temperature. As a result, accuracies are limited to no better than 1% with potential hysteresis problems. Advances in the use of metalized coatings may reduce this problem.

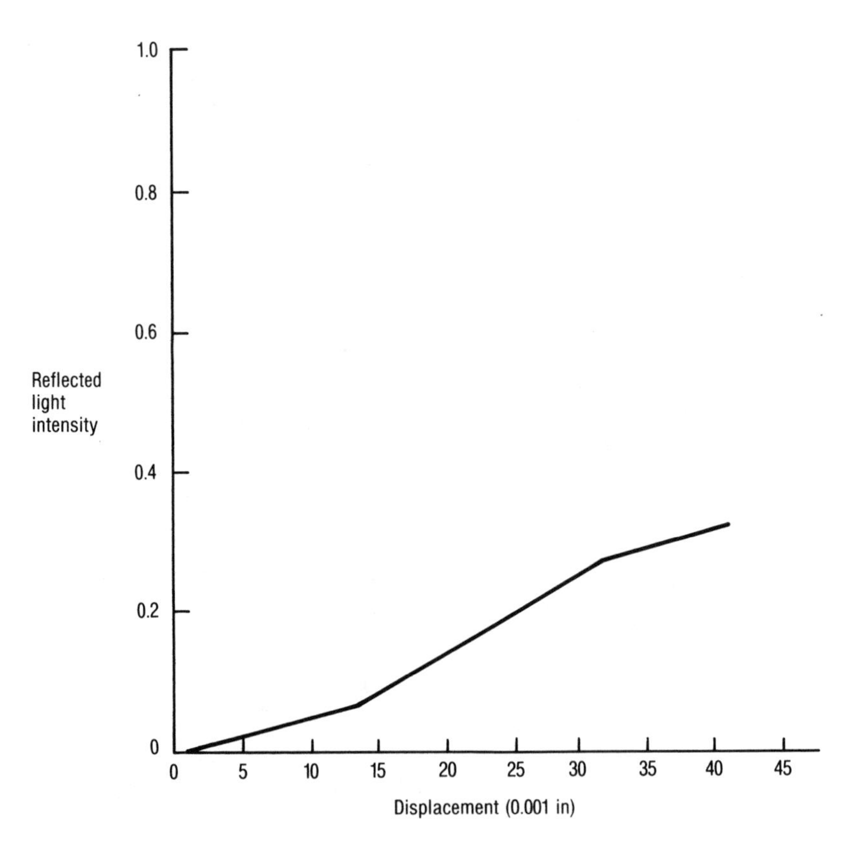

Figure 6-21
Transmission versus Displacement (Dark Field Microbending)

APPLICATIONS

Used as single-point sensors, reflective or microbending sensors have a wide range of applications. Figure 6-22 shows several applications using a reflective sensor. Measurement of axial motion, proximity, eccentricity, and shaft

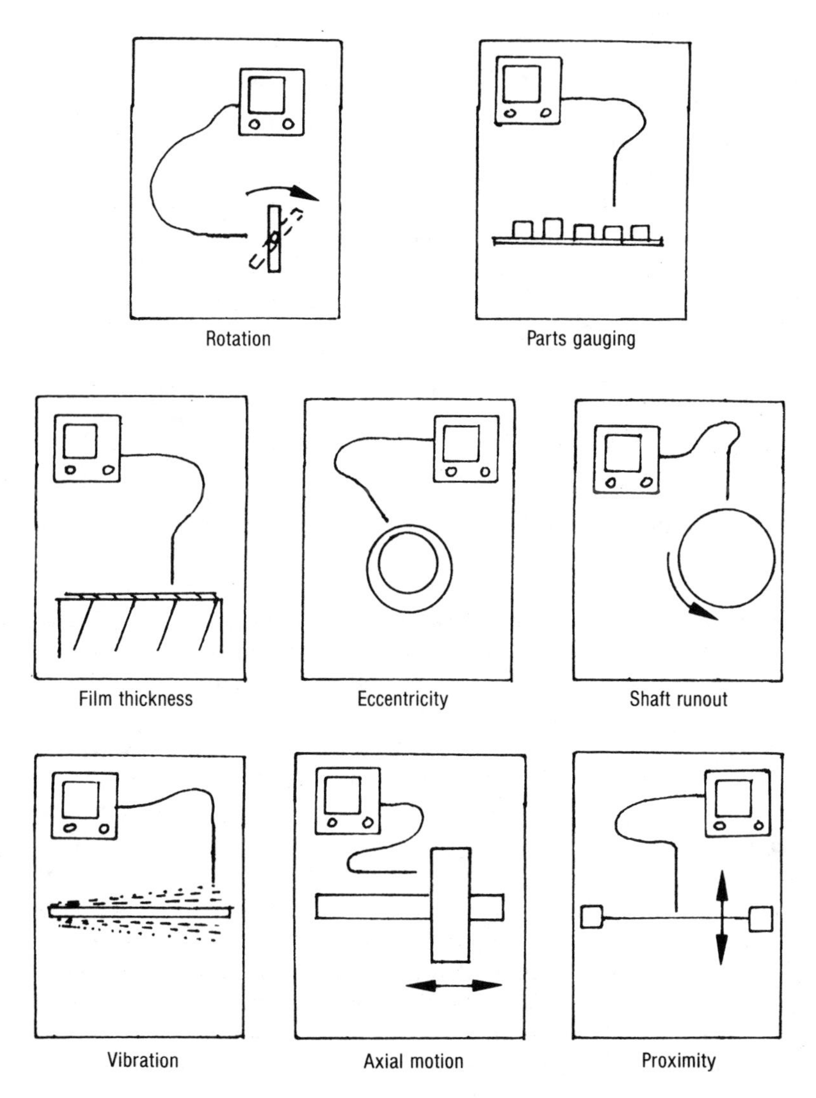

Figure 6-22
Displacement Sensor Applications

runout are straightforward displacement applications. Vibration measurement normally requires a high speed electro-optic interface to record the vibration "signature".

Film thickness measurement is somewhat more complex in that the reflective return signal from a target is perturbed by the presence of a film. The change in signal associated with the film can be correlated with its thickness. Parts gauging uses the analog displacement nature of the sensor output to discriminate high/low objects. The last application shown is for rotation. As indicated previously, displacement sensors are sensitive to rotation since the rotation causes the average target to probe distance-to-change as well as the angle of reflectance.

Another application is for dial encoding.[15] A specially fabricated cam is placed on a shaft that supports a meter dial, as shown in Figure 6-23. The cam increases the distance between the fiber optic probe and the reflective target in a step-wise sequence of 10 distinct distances. Each step corresponds to a digit on the 0–9 dial indicator. The linear portion of the back slope is used.

Microbending sensors have been used successfully as strain sensors. Two potential configurations are shown in Figure 6-24. In configuration (a) the fiber and the perturbing mechanism are mounted on the surface of the object to be measured. Only the perturbing mechanism is attached to the object, with the fiber free to bend. In configuration (b) the fiber is mounted much like a conventional strain gage. This configuration is not nearly as sensitive as the one using the perturbing mechanism, since the same displacement in strain will cause less bending. However, such a device would be good for large displacement applications, such as integration into composite structures to detect the possibility of catastrophic failure.

As mentioned previously, dual probes provide for greater sensitivity as well as directionality. Figure 6-25 depicts dual probe applications that include measurement of concentricity, diameter, alignment, thickness, and servo positioning. Consider the servo positioning application. Movement to the left or right a given distance corresponds to specific values for each probe. If the magnitude registered for one probe does not correspond to the calibrated magnitude registered for the other probe, a difficulty such as misalignment, uneven wear, or contamination could be signaled.

A rotational shaft with limited movement used in the measurement of torque is monitored by dual reflective fiber optic probes in Figure 6-26 and by dual microbending sensors in Figure 6-27. In both cases the shaft rotates an eccentric cam, causing displacement proportional to torque. Both sensors monitor magnitude and direction.

Accelerometers can be configured very much like the servo positioning sensor in Figure 6-25. The piston or pendulum on which the reflective

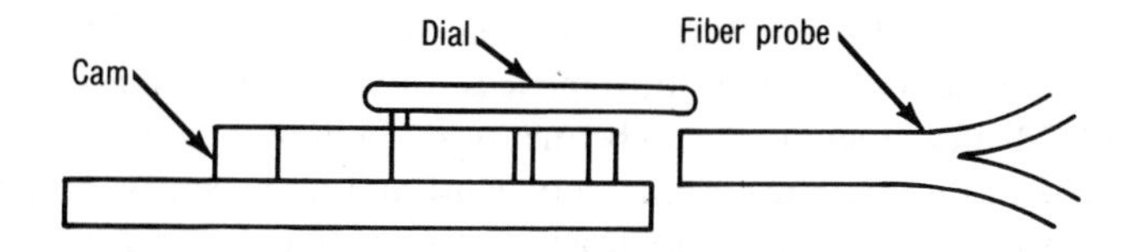

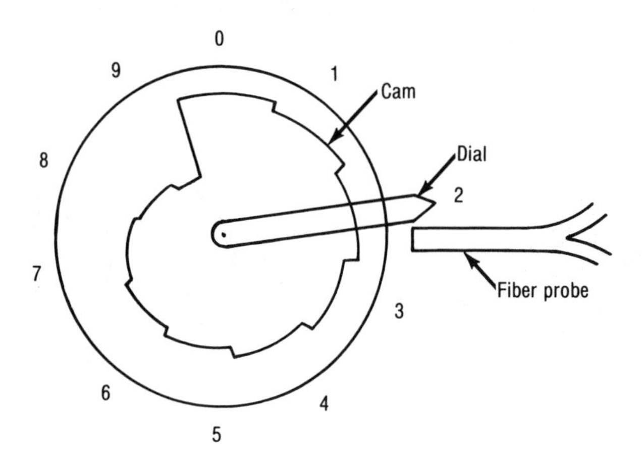

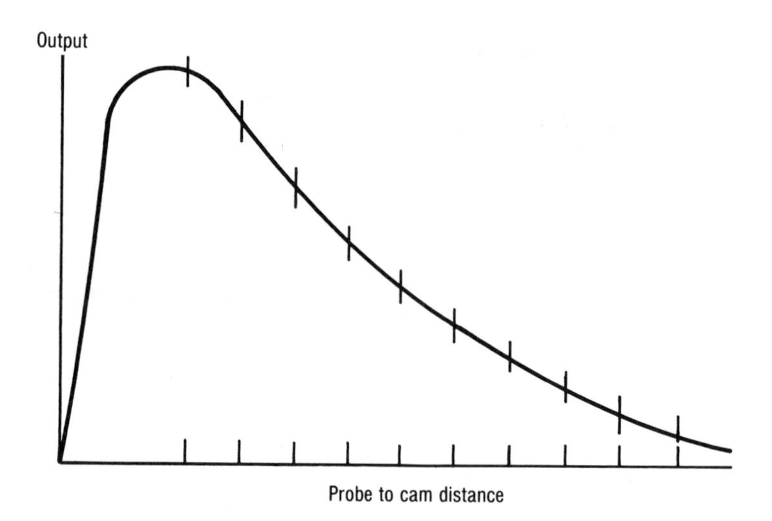

Figure 6-23
Dial Encoder Configuration

(reprinted by permission from SENSORS, © 1984 Helmers Publishing, Inc.)

targets are mounted is an inertial mass that is free to respond to acceleration.[17] The precise measured movement is a function of the acceleration.

Finally, displacement can also be measured in digital format using fiber optics. The sensor requires a gray code card and a transmission interrupt sensor of each bit of required accuracy. Typically, five or more transmissive sensors are required. The sensor concept eliminates intensity errors associated with analog sensors, but it is generally bulkier, more costly, and problematic in dirty environments.

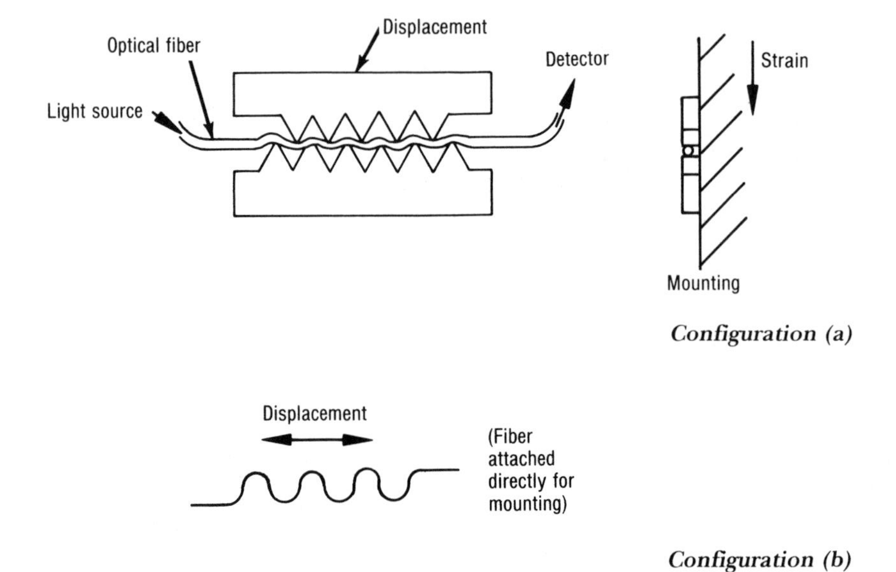

Configuration (a)

Configuration (b)

Figure 6-24
Microbending Strain Sensors

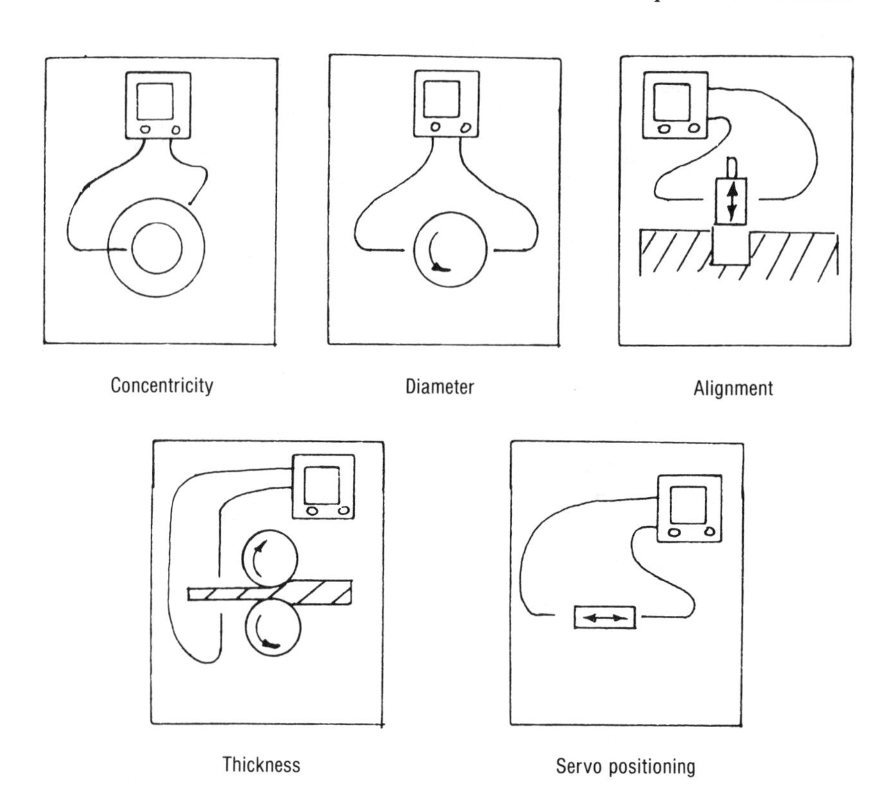

Figure 6-25
Typical Applications — Dual Probe

REFERENCES

1. Krohn, D. A., October 1984, "Fiber Optic Displacement Sensors," *Proceedings of the ISA — Houston, TX*, Vol. 39, Part 1, pp 331–40.

2. Kissinger, C. D., and Howland, B., Feb. 24, 1976, "Fiber Optic Displacement Measuring Apparatus," U.S. Patent 3,940,608.

3. Bejczy, A. K., Primus, H. C., and Herman, W. A., March 1980, "Fiber Optic Proximity Sensor," *NASA Tech. Brief*, Vol. 4, No. 3, Item 63, JPL Report NPO-14653/30-4279.

4. Coulombe, R. F., December 1984, "Fiber Optic Sensors Catching Up with the 1980's," *Sensors*, pp 5–10.

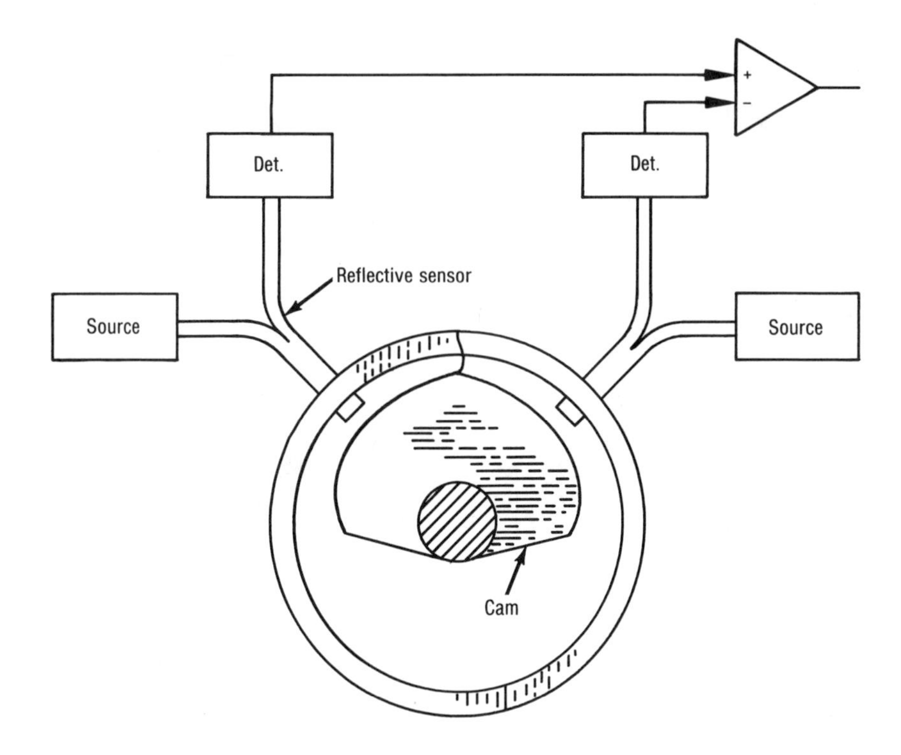

Figure 6-26
Torque Sensor Using Dual Reflective Sensors

5. Rasmay, M. M., Hockman, G. A., and Kao, K. C., "Propagation in Optical Fiber Waveguides," *Electrical Communication*, Vol. 50, No. 3.
6. Krohn, D. A., 1982, "Fiber Optic Sensors: A Technology Overview," *Innovations*, pp 11–13.
7. Lugakos, N., Trott, W. S., Hickman, T. R., Cole, J. H., and Bucaro, J., October 1982, "Microbend Fiber-Optic Sensor as Extended Hydrophone," *IEEE J. of Quantum Elctronics*, Vol. QE-18, No. 10, pp 1633–1638.
8. Giallorenzi, T. G., et al., April 1982, "Optical Fiber Sensor Technology," *IEEE J. of Quantum Electronics*, Vol. QE-18, No. 4, p. 665.
9. Marcuse, D., 1976, "Microbending Losses of Single-Mode, Step-Index and Multimode, Parabolic-Index Fibers," *Bell System Technical Journal*, Vol. 55, No. 7, pp 937–955.

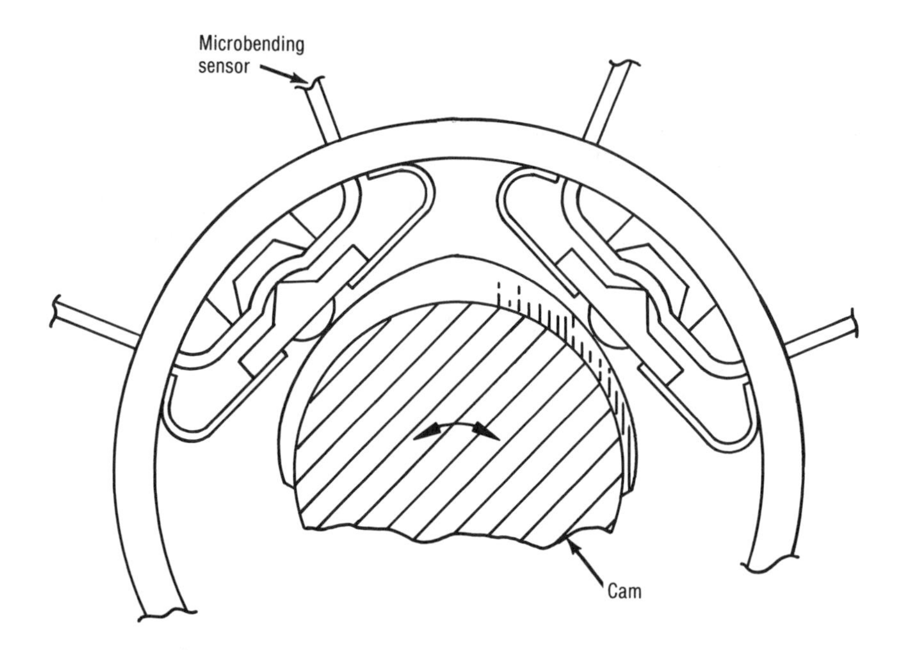

Figure 6-27
Torque Sensor Using Dual Microbending Sensors

10. De Formal, F., Arnaud. J., and Facq, P., 1983, "Microbending Effects on Monomode Light Propagation in Multimode Fiber," *J. Opt. Soc. Am.*, Vol. 73, No. 5, pp 661–668.

11. Davis, C. M., et al., 1982, *Fiber Optic Sensor Handbook*, Dynamic Systems, Reston, Virginia.

12. Lagakos N., Litovitz, T., Macedo, P., Mohr, R., and Meister, R., 1981, "Multimode Optic Fiber Displacement Sensor," *Applied Optics*, Vol. 20, No. 2, pp 167–168.

13. Fields, J. N., Asawa, C. K., Ramer, O. G., and Barnoski, M. K., 1980, "Fiber Optic Pressure Sensor," *J. Acoust. Soc. Amer.*, Vol. 67, pp 816–818.

14. Krohn, D. A., 1983, "Fiber Optic Sensors in Industrial Applications: An Update," *Proceedings of the ISA — Houston, TX*, Vol. 38, pp 877–890.

15. Krohn, D. A., Buffone, L. A., and Vinarub, E. I., 1985, "Fiber Optic Dial Encoder," U.S. Patent 4,500,870.

16. Jones, R., July 1984, "Optical Techniques for Inspection and Sensing," *Sensor Review*, pp 116–119.

17. Soref, R. A., and McMahon, D. H., 1984, "Tilting-Mirror Fiber Optic Accelerometer," *Applied Optics*, Vol. 23, No. 3, pp 486–491.

Temperature Sensors

7

INTRODUCTION

Several considerations drive the need for fiber optic temperature sensors. Sensors are needed to operate in strong electromagnetic fields. Sensors with metallic leads will experience eddy currents in such environments, which will create both noise and the potential for heating of the sensor, which, in turn, causes inaccuracy in the temperature measurement. Fiber optic temperature sensors that do not use metallic transducers allow for minimized heat dissipation by conduction and provide quick response. Since they are less perturbing to the environment, they have the potential for extreme accuracy.(1)

Several fiber optic sensing concepts have been applied to temperature measurement; reflective, microbending, intrinsic, as well as other unique intensity-modulated approaches are described in the literature.(2,3,4) Phase modulated concepts also have been applied to temperature sensing.

REFLECTIVE CONCEPTS

As described in Chapter 6 on displacement, reflective sensors can very accurately determine position. Figure 7-1 depicts a bimetallic element attached as a transducer to a bifurcated reflective fiber optic probe. The bimetallic element can be designed to provide a snap action at a specific temperature, moving abruptly relative to the probe tip, thus resulting in a switching action at a set temperature point. The element can also be designed to move continuously and provide movement proportional to temperature in an analog mode.

A response curve for a simple analog bimetallic element is shown in Figure 7-2.[5] An actual sensor is shown in Figure 7-3. Figure 7-4 shows another approach using differential thermal expansion. The two targets (alumina and silica), which have substantially different thermal expansions, reflect light into the output optical fiber. As the targets move relative to each other as a function of temperature, interference fringes are created and the fringes are counted. The concept provides a digital indication of temperature using reflectance.[6]

As mentioned previously, active sensing materials can be placed in the optical path of a reflective probe to provide an enhanced sensing function. A wide variety of materials can be used for temperature sensing including liquid crystals, birefringent materials, semiconductor materials, materials that fluoresce, and materials that can change spectral response other than fluorescence.

Liquid crystals exhibit color changes and/or reflectivity changes at a fixed wavelength due to temperature changes. The sensor mechanism is quite simple. Light passes through the input fiber and is reflected from the liquid crystal. The reflected light is detected, the intensity of which is a function of temperature. The working range is quite limited, 35 °C to 50 °C, but accuracies as good as 0.1 °C have been reported. This concept is limited to monitoring biological processes.[1]

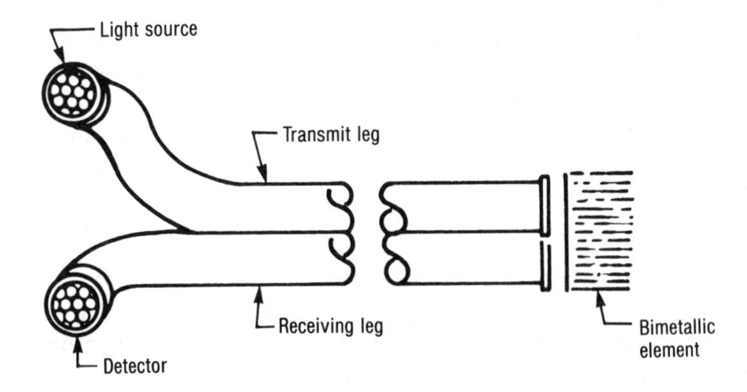

Figure 7-1
Reflective Fiber Optic Temperature Sensor Using a Bimetallic Transducer

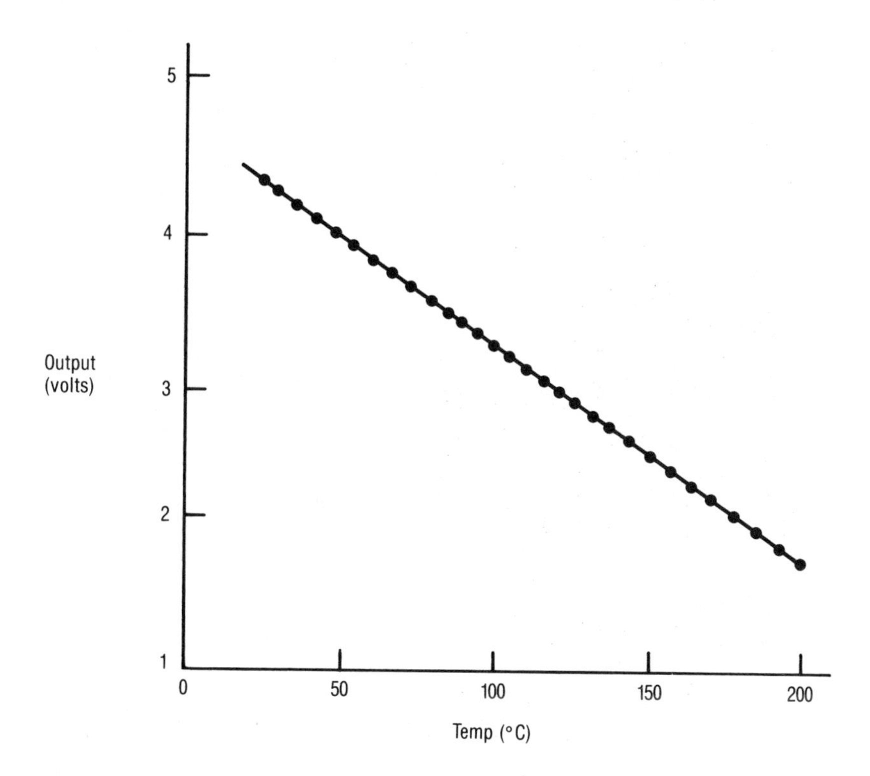

Figure 7-2
Response Curve for a Reflective Fiber Optic Temperature Sensor Using a Bimetallic Transducer

Figure 7-5 shows a reflective temperature sensor that uses a birefringent crystal. Birefringent crystals are optically transparent crystalline materials in which the indices of refraction are different for the orthogonally polarized light waves. The light is carried to the crystal by a transmitting fiber and passes through a polarizer prior to passing through the crystal. The light is reflected from a mirror and passes back through the crystal, polarizer, and receiving fiber to the detector. There are materials in which the index of refraction and therefore the birefringence is a strong function of temperature. The net change in birefringence alters the intensity of received light, which is proportional to temperature.(1) The sensor is limited to biological process temperature ranges.

Another approach uses a semiconductor crystal as a target.[1] Temperature measurement is based on the fact that band edge absorption of infrared light in crystals such as GaAs is temperature dependent. Figure 7-6 depicts the sensor configuration. As with the liquid crystal, the sensor is accurate but limited to values below 47°C.

Figure 7-3
Actual Reflective Fiber Optic Temperature Sensor Using a Bimetallic Transducer

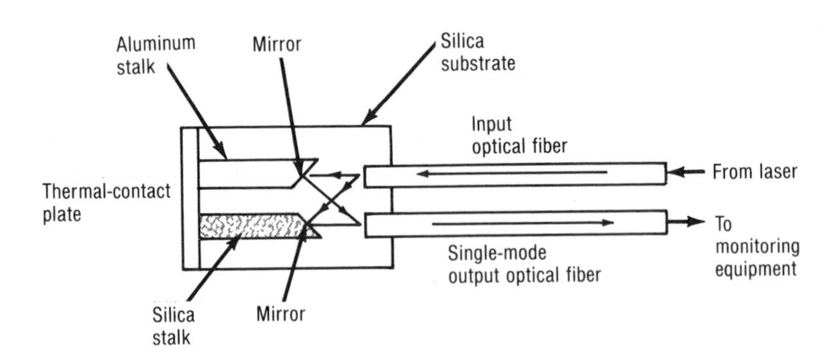

Figure 7-4
Reflective Fiber Optic Temperature Sensor Using Differential Thermal Expansion

Another reflective concept is spectral modulation.[7] The configuration is shown in Figure 7-7. The sensing target element acts as a spectral mirror, changing the spectral reflectance over the light source bandwidth. The bandwidth of the light source is sufficiently broad that the sensing element can resolve the wavelength into discrete components. The ratio of two of the discrete component wavelengths varies in a manner proportional to the perturbing environment — in this case, temperature. The target for a temperature sensor is comprised of a material with a high refractive index change with temperature. Light is transmitted through the target to a stationary reflector and back through the target material. Unlike a bimetallic

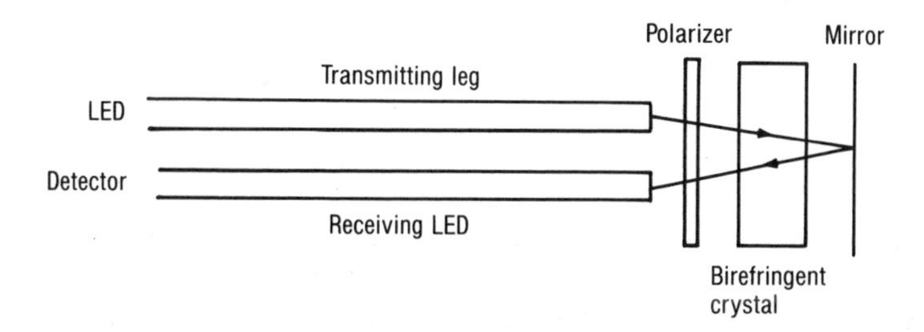

Figure 7-5
Reflective Fiber Optic Temperature Sensor Using a Birefringent Crystal

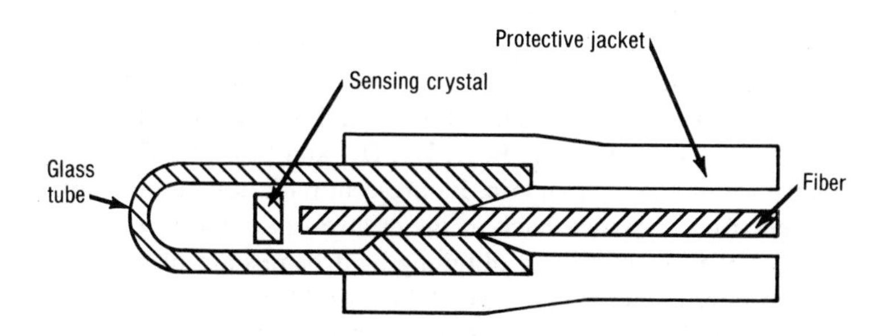

Figure 7-6
Reflective Temperature Probe in Which a Semiconductor Crystal is Hermetically Sealed Inside a Glass Capillary Tube

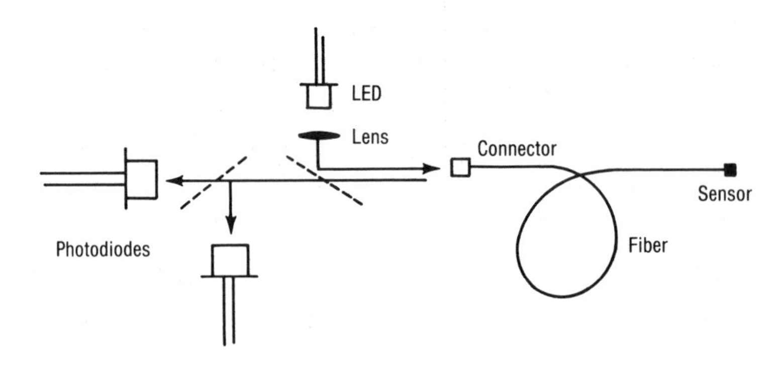

Figure 7-7
Schematic Representation of a Reflective Fiber Optic Sensor Using Spectral Modulation

diaphragm there is no movement; the pathlength, and hence the spectral response, changes due to index change. The target material is typically 1 to 2 wavelengths deep. The small size of the sensor allows it to respond quite rapidly to dynamic temperature changes. The ratiometric signal versus temperature is shown for the sensor in Figure 7-8. The upper limit of the sensor is dictated by the fiber coatings. The resolution is approximately 1 °C over a linear range of 400 °C.

The reflective concept can be expanded to include fluorescent targets.[8,9] In essence, the target does not reflect but absorbs the incident light and emits fluorescent radiation. The sensor uses a single fiber made of pure quartz to maximize the transmission of input UV light. The sensor tip is shown in Figure 7-9. The fiber tip is coated with a phosphor layer and encapsulated. The sensor system is shown in Figure 7-10. An ultraviolet lamp is the light source. The light is filtered and focused onto the fiber with a series of optical elements (L_1, F, D, D_2, D_3, and L_2). The UV light excites the phosphor on the fiber end tip. Fluorescent radiation in the visible region of the spectrum is carried back to the electro-optic interface via the same fiber. Since the incoming and outgoing light beams are of different wavelength, no interference takes place. A beam splitter is used to separate two beams. The two beams are isolated by interference filters (IF_1 and IF_2). The detected intensity of each beam is determined and fed to a microprocessor, which calculates the intensity ratio, which is a direct function of the temperature of the phosphor.

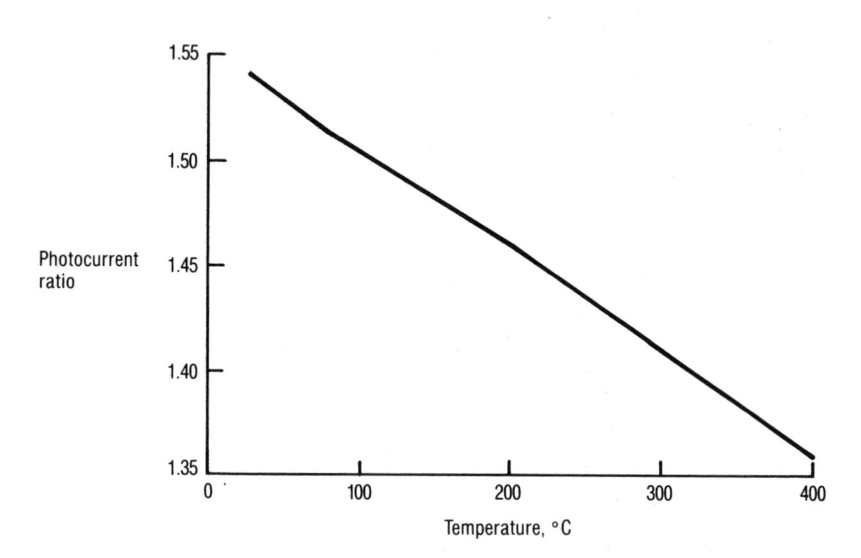

Figure 7-8
Ratiometric Signal for Typical Temperature Sensor Using Spectral Modulation

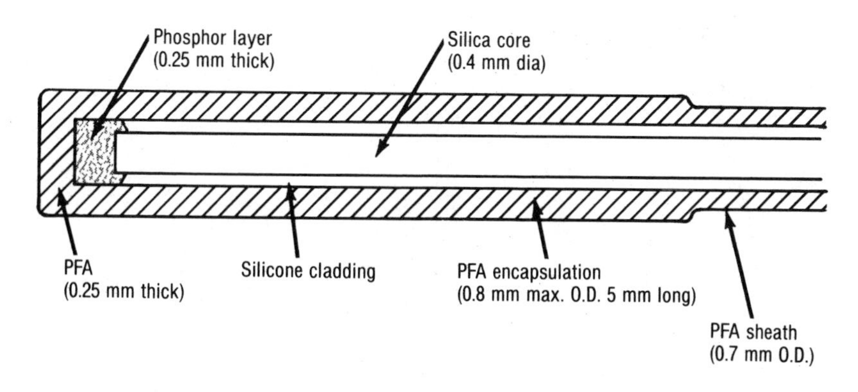

Figure 7-9
Sensor Tip (Highly Enlarged) for a Fluorescent Sensor

(reprinted by permission from *Biomedical Thermology*, © 1982 Alan R. Liss, Inc.)

To be more specific, the principal operation of a fluorescent temperature sensor is to measure the relative intensities of two sharp fluorescent emission lines that are given off when the phosphor is excited by UV radiation. The lines vary differently with temperature; as a result their intensity ratio is a measurement of temperature.

A phosphor that has been used successfully in a fluorescent sensor is europium-activated gadolinium oxysulfide.[8] The excitation spectrum and emission spectrum are shown in Figure 7-11. The emission lines are temperature dependent. They are denoted by L, M, or H (low, medium, or high temperature), which defines the temperature range in which the lines quench or fade. The intensity ratio of lines marked a and c have been used to determine temperature.

The operating range of the sensor is − 50 °C to above 250 °C. Over this broad temperature range, the accuracy is on the order of 1 °C. Over a much narrower temperature range, the accuracy approaches 0.1 °C. An important feature of this sensing concept is that the measured temperature is a direct

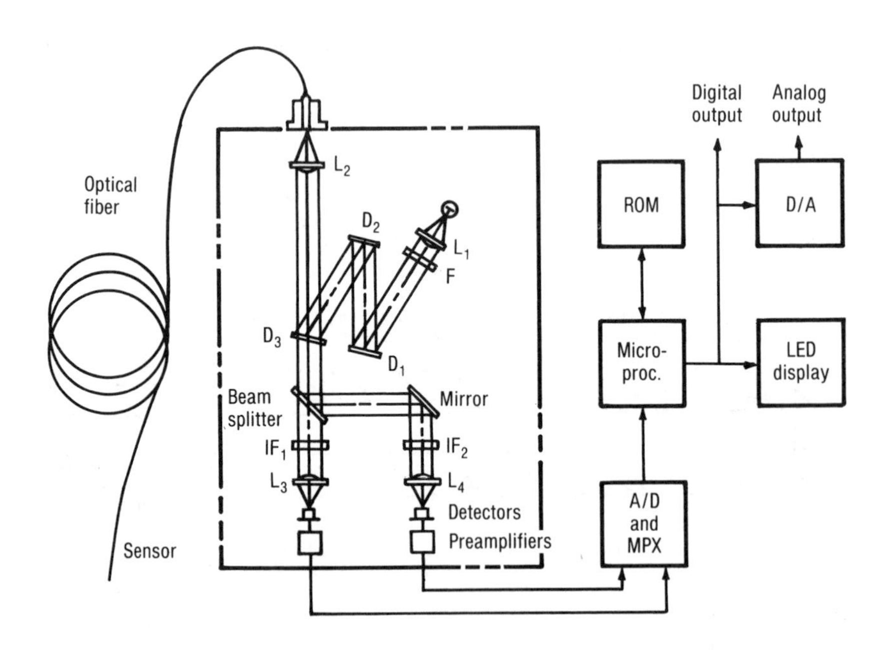

Figure 7-10
Schematic Representation of the Fluorescent Temperature Sensor System

(reprinted by permission from *Biomedical Thermology*, © 1982 Alan R. Liss, Inc.)

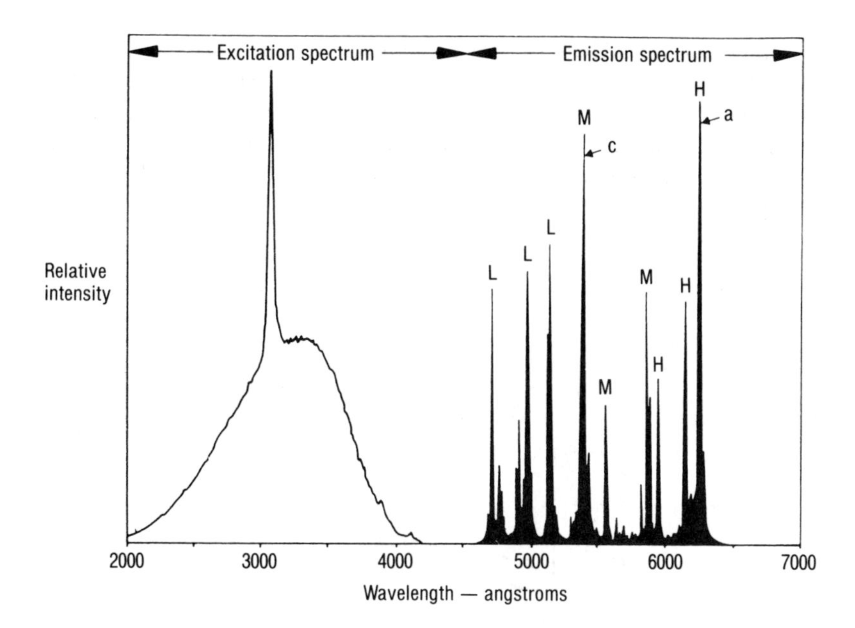

Figure 7-11
Excitation and Emission Spectrum for a Fluorescent Fiber Optic Temperature Sensor

(reprinted by permission from *Biomedical Thermology*, © 1982 Alan R. Liss, Inc.)

function of phosphor temperature and is independent of variations caused by the light source, microbending in the leads, and connector-related intensity variations.

MICROBENDING

Microbending can be used as a temperature monitor in two different schemes.[4] The sensor can function by a change in the refractive index difference or by a change in the radius of bending with temperature. As described in Chapter 3, the microbending loss is a function of the difference in the refractive index between the core and cladding, Δn. For a given degree of bending as Δn becomes smaller, the fiber becomes more lossy and leaks more light when the fiber is being bent. The temperature effects are pronounced on polymer-clad glass fibers where the temperature dependence of the refractive index for the polymeric coating is significantly larger than for the glass core. Therefore, such fibers are sensor candidates. Glass-clad glass fibers can also exhibit such behavior. Gottlieb and Brandt(4) have characterized several commercial glass combinations in which the indices of refraction cross over at a given temperature, as shown in Figure 7-12. These fibers have the potential to act as temperature switches at the crossover point.

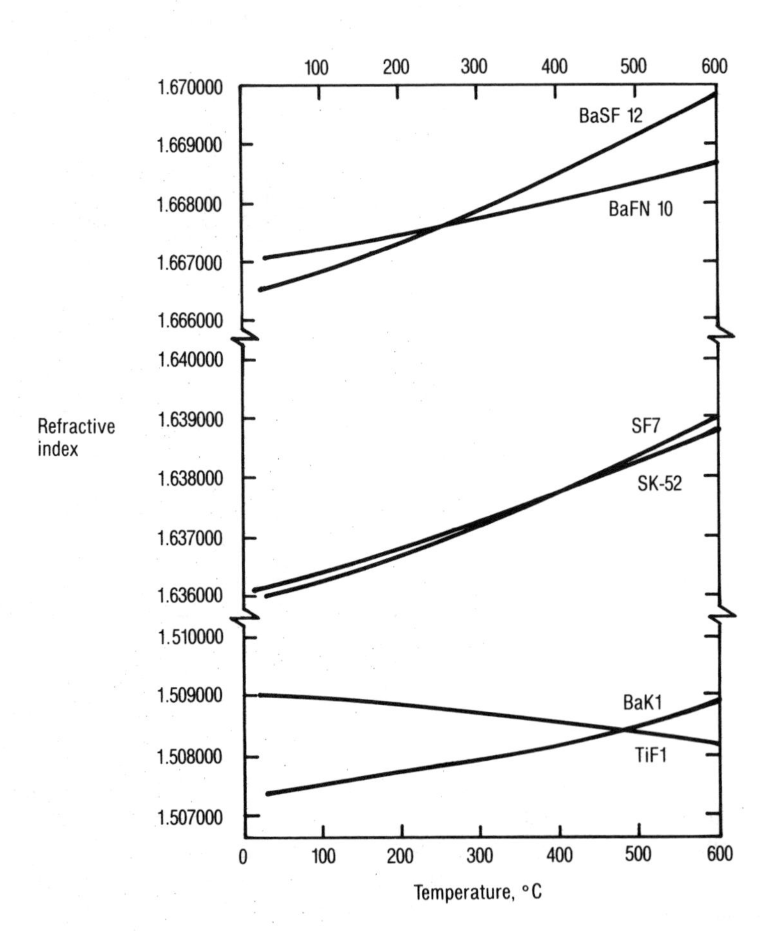

Figure 7-12
Crossover of Refractive Index of Several Pairs of Fiber Glasses. The Codes Refer to Schott Glass Designations for Various Commercial Glasses.(4)

The change in bend radius with temperature can easily provide a sensing function by using a bending mechanism device with a high thermal expansion. The dependence of bending loss on radius of curvature is shown in Figure 7-13 for several fibers.[4] It is clearly shown that relatively small radius changes can significantly increase loss. If a continuous multipoint sensor is being designed, the parameters of the fiber and microbending transducer must be chosen so that one hot spot does not shut the entire fiber off. Figure 7-14 shows various configurations of a microbending temperature sensor. In Figure 7-14(a), the fiber is wrapped around a pipe that can expand or contract; the sensor is monitoring the pipe in a continuous manner. In Figure 7-14(b) the configuration shows a point monitor where only the specific sensing point has sufficient thermal movement to affect the sensor.[4] In Figure 7-14(c), the configuration is similar to a strain gage; if a microbending-sensitive fiber is bonded to a high thermal expansion material in the configuration shown, lateral expansion or contraction will change the curvature of the fiber.

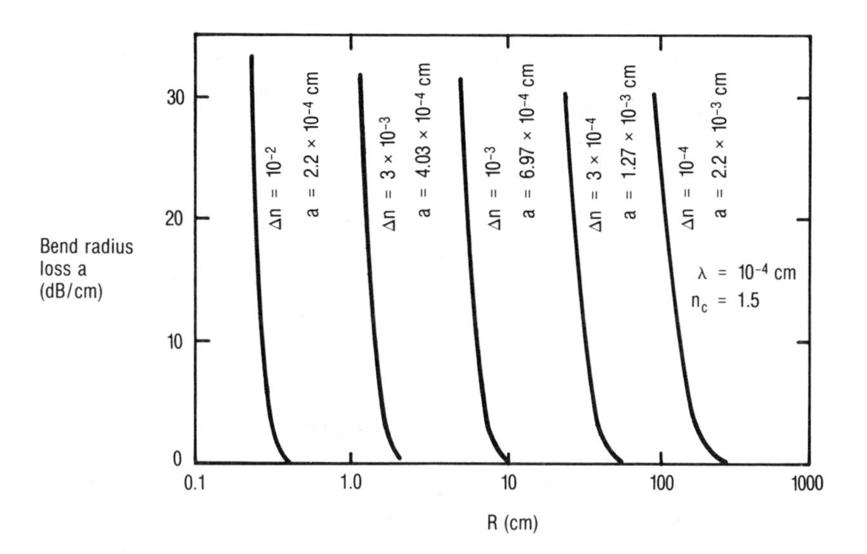

Figure 7-13
Dependence of Bending Loss on Radius for Several Fibers.[4]

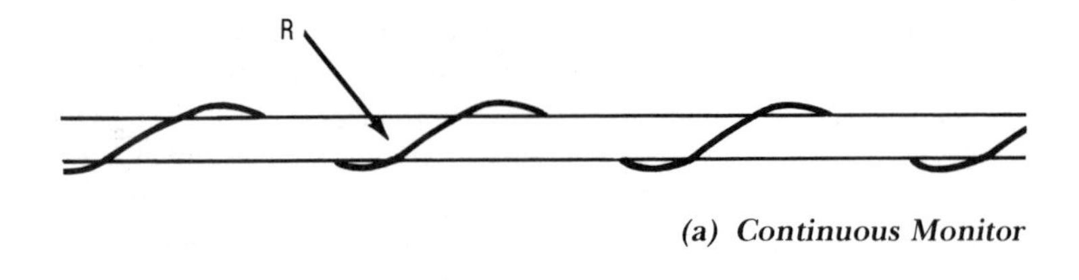

(a) Continuous Monitor

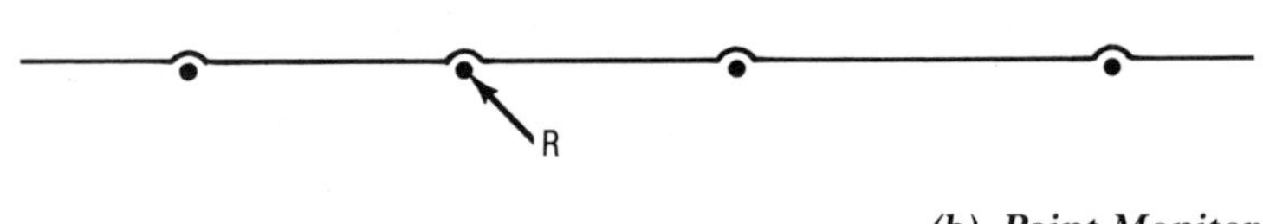

(b) Point Monitor

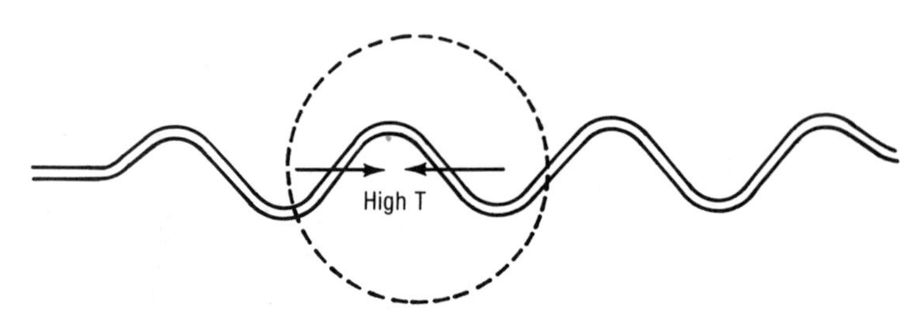

(c) Variable Radius Monitor

Figure 7-14
Several Configurations for a Bend Loss Temperature Monitor[4]

OTHER CONCEPTS

Blackbody fiber optic temperature sensors are based on the fact that thermal radiation is emitted when a material is heated.[2,3,10] The intensity and wavelength of the radiation is a function of the temperature. The wavelength of radiation is also a function of temperature. At low temperatures ($<100\,^{\circ}C$) the wavelength of transmission is above 4 microns so that most conventional fibers are not effective. Typically, pure silica fibers have a transmission cutoff near 2 microns. At higher temperatures, the wavelength requirement for sensing drops into the operational range of silica fiber.

The most advanced temperature sensors using this approach have optical fiber made of sapphire. A typical configuration is shown in Figure 7-15.[2] The tip of the fiber is coated with a thin film of a noble metal to form a blackbody cavity. The sensor has an operational measurement range of 500 °C to 2000 °C. The response time is potentially several orders of magnitude faster than for a conventional thermocouple.

Another approach to a temperature sensor involves coupling between parallel waveguides. Chapter 2 discussed the possibility of cross coupling energy from one fiber to another. The parameters that affect the cross coupling of light energy include the spacing between the parallel waveguides, the refractive index of the material between them, and the length over which they run parallel. Temperature monitoring can occur if the spacing is affected by thermal expansion. In such a sensor, as the temperature is raised, the spacing is decreased and, therefore, cross coupling takes place. The magnitude of cross coupling light is proportional to temperature. A similar effect can be achieved if the index of refraction of the material between the waveguides decreases with increasing temperature. Devices of this type can be configurated as point sensors or continuous area monitors.

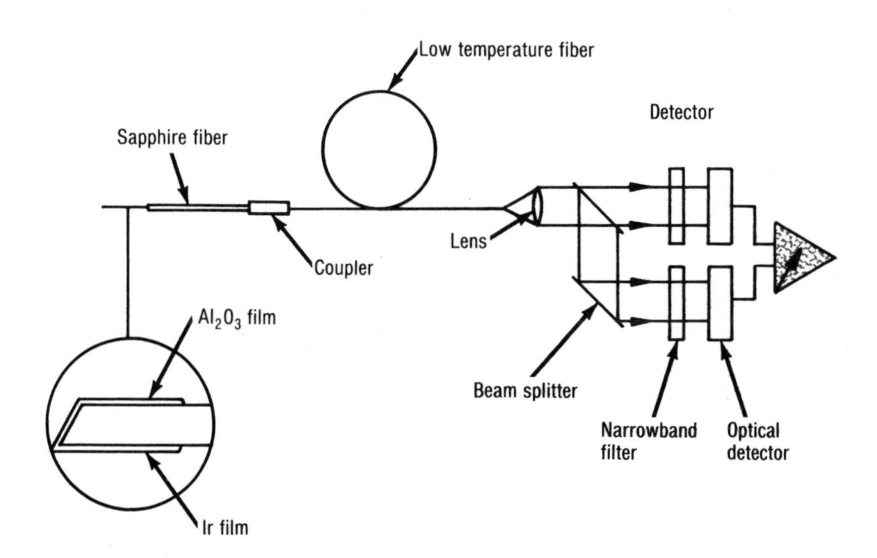

Figure 7-15
Blackbody Temperature Sensor Configuration.[2]

INTRINSIC CONCEPTS

An intrinsic sensor is defined as one in which the sensing function takes place in the fiber itself (core, cladding, or coating) and the intensity of light transmitted in the fiber is proportional to the perturbing environment. An intrinsic sensor for temperature involves the phenomenon of absorption. It has been found that rare earth materials such as neodymium (Nd) and europium (Eu), when added to a conventional glass, result in an absorption spectra with temperature-sensitive properties.[11,12] Two wavelengths were found with unique temperature behavior for Nd-doped fibers.

As shown in Figure 7-16, at 840 nm the absorption decreases with temperature; at 860 nm the reverse is true up to 500 °C. The intensity of each of the two wavelengths is determined and the ratio provides a measure of temperature, as shown in Figure 7-17. The operational range of the sensor is about 0 to 800 °C. A schematic of the sensor system is shown in Figure 7-18.

It is important to note that the microbending temperature sensor described previously is an intrinsic sensor using the mechanism of refractive index change. An extremely important feature of intrinsic sensors involves

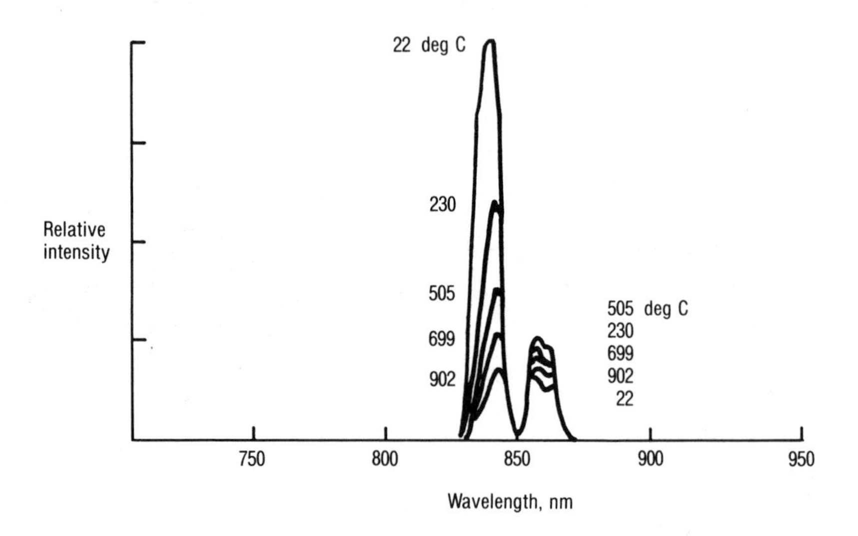

Figure 7-16
Temperature Sensitivity Absorption Spectra for a Neodymium Doped Glass Fiber

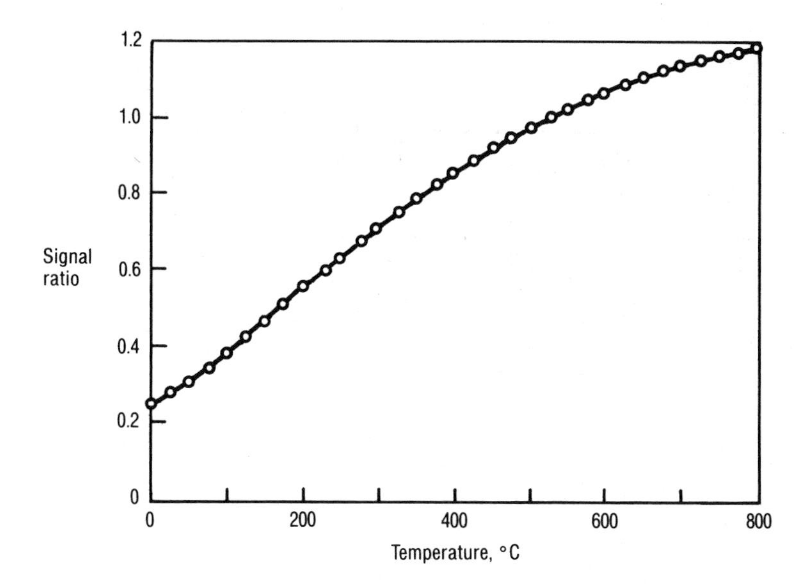

Figure 7-17
Temperature Response Curve for Neodymium-Doped Glass Fiber
(© 1983 IEEE; reprinted with permission)

the activation mechanism. Typically, if the intrinsic sensing behavior is occurring in the cladding or coating, microbending or macrobending must be used to cause an interaction with the light transmitted in the core.

INTERFEROMETRIC CONCEPT

As was discussed in Chapter 4, interferometric sensors are sensitive to both changes in length and refractive index. If the parameters in the sensing fiber change relative to the reference fiber, a phase shift occurs. For an interferometric temperature sensor, the phase shift as a function temperature, $\frac{\Delta\phi}{\Delta T}$ is given by:[13]

$$\frac{\Delta\phi}{\Delta T} = \frac{2\pi L}{\lambda} \frac{n\Delta L}{L\Delta T} + \frac{\Delta n}{\Delta T} \qquad (7\text{-}1)$$

where L is fiber length and n is refractive index. For high silica fibers, the

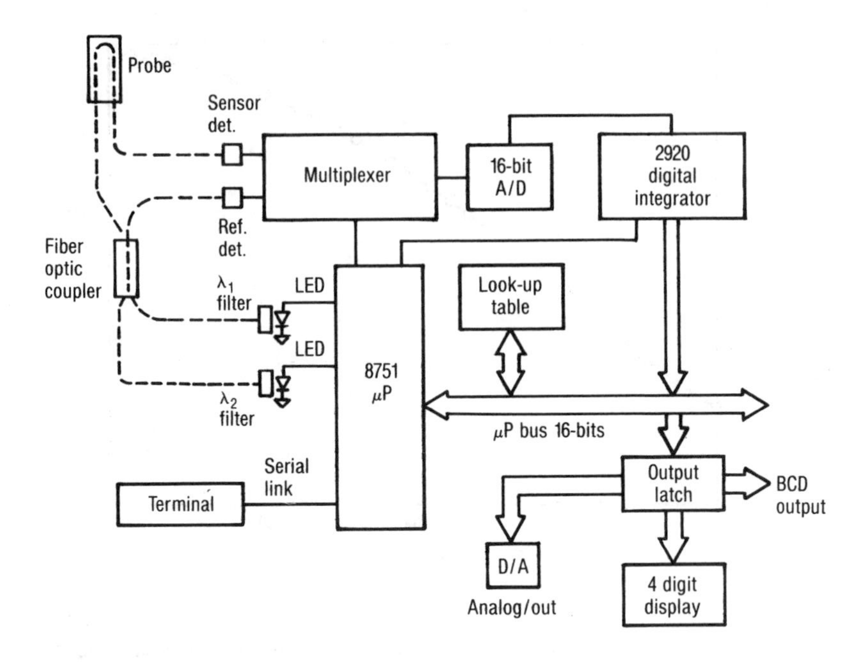

Figure 7-18
Sensor System Schematic

(© 1983 IEEE; reprinted with permission)

$\Delta n / \Delta T$ term dominates, since the length is relatively insensitive to temperature due to the low coefficient of expansion of silica. The sensitivity of this approach was calculated to have values of 10^{-8}°C. Generally, such high sensitivities are unnecessary, and less expensive techniques will be used.

In new integrated optical techniques the interferometer is deposited on a ceramic chip, as shown in Figure 7-19.[14] A beam of laser light is split and passes along each path. The velocity of light is a function of the refractive index of the material, which, in turn, is a function of temperature. As the temperature varies, the two beams will be recombined, but out of phase. The interference will create an intensity that varies sinusoidally with a period proportional to the difference in the length of the two paths. Increasing the difference in path lengths increases sensitivity; by using several parallel interferometers, the dynamic range can be boosted.

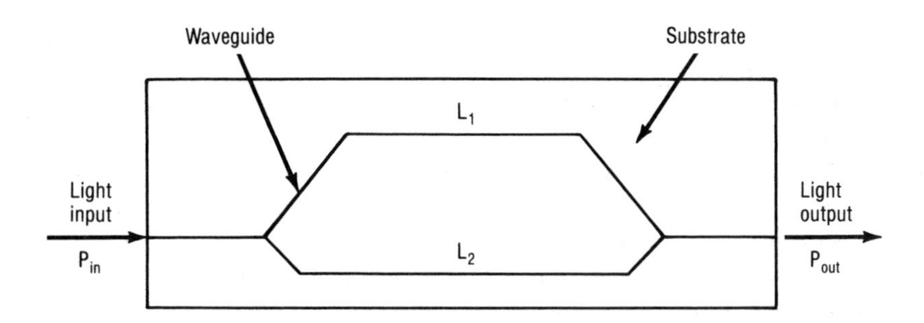

Figure 7-19
Schematic Diagram of Interferometer Mounted on a Ceramic Chip with Unequal Arm Lengths L_1 and L_2

(reprinted by permission from *The MIT Report*, © 1981; Massachusetts Institute of Technology)

APPLICATIONS

The range of application of temperature sensors is quite broad: from monitoring relatively low temperature biological processes to monitoring high temperature engine parts. The three concepts that have had the greatest use include the reflective-bimetallic sensor, the fluorescent sensor, and the blackbody sensor.

Bimetallic sensors are available in a range from 5°C to 300°C. The potential for use lies in applications such as engine control, air compressors, and industrial processing equipment. Such sensors can be made rugged to withstand shock and vibration. Generally, they are large relative to the fiber and intrinsic sensors and as a result have limited dynamic response.

The fluorescent sensor has no metallic components and can be put into an extremely small package. It is especially useful in applications requiring electrical immunity or isolation. The high accuracy (0.1°C) and small size have made the sensor concept ideal for biological and physiological applications. Hyperthermia, one such area, is a potential cancer therapy requiring local tissue heating using an rf field. The fluorescent sensor has been successfully used in such applications. Any relatively low temperature (less than 250°C) processes using rf heating (such as wood processing and revulcanization of rubber) are potential candidates for this sensor type. The technique is also useful in diagnostic instrumentation associated with high voltage equipment. The detection of localized heating can often predict the onset of equipment malfunction.

The blackbody sensor concept has a range of 500–2000°C. Due to its small size and extremely fast response time, it has application in many high temperature processes such as characterizing the heat flow in a gas turbine engine.

The intrinsic concept (such as the approach using absorption) provides a sensing mechanism that can detect hot spots. If such an intrinsically sensing fiber were used in electric motor or transformer windings, it could provide an overheat alarm. It also has potential in distributive temperature sensors, which will be discussed in a later chapter.

REFERENCES

1. Wickersheim, K. A., and Alves, R. B., December 1979, "Recent Advances in Optical Temperature Measurement," *Industrial Research/Development*.

2. Dils, J. J., 1973, "Optical Fiber Thermometry, *J. Appl. Phys.*, Vol. 84, p. 1198.

3. Gottlieb, M., and Brandt, G. B., 1981, "Fiber Optic Temperature Sensor Based on Internally Generated Thermal Radiation," *Applied Optics*, pp 3408–3414.

4. Gottlieb, M., and Brandt, G. B., 1979, "Measurement of Temperature with Optical Fibers Using Transmission Intensity Effects," *Proceedings*, Electro-Optics Conference, Anaheim.

5. Krohn, D. A., 1983, "Fiber Optic Sensors in Industrial Applications — An Update," *Proceedings of the ISA*, Houston, TX, pp 877–890.

6. Anonymous, 1984, "Multipurpose Photonic Transducers," *NASA Tech. Briefs*, pp 520–521.

7. Saaski, E. W., Hartl, J. C., Mitchell G. L., 1986, A Fiber Optic Sensing System Based on Spectral Modulation," *Proceedings of the ISA*, Houston TX.

8. Wickersheim, K. A., and Alves, R. V., 1982 "Fluoroptic Thermometry: A New Immune Technology," *Biomedical Thermology*, Alan R. Liss, New York, pp 547–554.

9. Alves, R. V., Christol, J., Sun, M., and Wickersheim, K. A., 1983, "Fluoroptic Thermometry: Temperature Sensing Using Optical Fibers," *Proceedings of the ISA*, Houston TX, pp 925–931.

10. Dils, R. R., and More, M. P., 1986, "Optical Fiber Thermometer Measurements in Automotive Engineers," *Proceedings of the ISA*, Houston, TX.

11. Snitzer, E., Morey, W. W., and Glenn, W. H., 1983, "Fiber Optic Rare Earth Temperature Sensors," *Proceedings IEEE/OFS*, Vol. 221, p. 79.

12. Grattan, K. T. V. and Palmer, A. W., 1985, "Simple Neodymium Rod Fiber-Optic Temperature Sensor," *Techincal Digest OFC/OFS*, p. 142.

13. Giallorenzi, T. G., Bucaro, J. A., Dandridge, A., Sigel, G. H., Cole, J. H., and Rashleigh, S. C., 1982, "Optical Fiber Sensor Technology," *IEEE J. of Quant. Elec.*, Vol. QE-18, No. 4, pp 626–665.

14. Johnson, L. M., Leongerger, F. J., and Pratt, G. W., November 1981, "Integrated Optical Technologies for Sensing and Signal Processing," Discussed in Research Notes, New Sensor Technologies, *The MIT Report, MIT*, Cambridge Mass., pp 1–3.

Pressure Sensors

8

INTRODUCTION

An excellent review of pressure monitoring devices and principles can be found in References 1 and 2. In several concepts for measuring pressure the sensor actually measures displacement, such as for devices that incorporate diaphragms, bellows, or bourdon tubes. Therefore, the fiber optic intensity-modulated sensing approaches are applicable to pressure measurement. These approaches include transmission, reflection, and microbending.

The driving force for the use of fiber optic pressure sensors has been the small size, freedom from EMI and RFI, accuracy, and, in the case of the reflective displacement sensor, noncontact. The potential small size of a reflective diaphragm sensor has created considerable interest in medical applications. Both intensity-modulated and phase-modulated fiber optic pressure sensors are now being designed for industrial use.

The interferometric approach has been most widely used in pressure sensing due to the fact that the fiber itself can act as the transducer. The extreme accuracy and wide frequency response have made interferometric sensors especially attractive as acoustic pressure sensors.

TRANSMISSIVE CONCEPT

Transmissive fiber optic pressure sensors have been divided into two basic categories. In the first category, the transmitting and receiving fiber legs remain fixed, and the modulation occurs by an object partially obscuring the light path.[3,4] In the second category, the fibers can move relative to each other to provide the modulation.

Figure 8-1 illustrates a transmissive sensor in which a shutter interrupts the light path in a manner proportional to pressure intensity. Using a reference and sensing channel provides ratiometric data with achievable full scale accuracies of 0.1%.

Figure 8-2[5] shows a transmissive sensor in which a grating is placed in the optical path. One grating is fixed while the other moves due to pressure. As the grating spacing becomes smaller, the sensitivity increases, providing the most sensitive transmissive pressure intensity-modulated sensor reported.

The concept of frustrated total internal reflection, which is a modified transmissive concept, is decribed in Chapter 3, "Intensity-Modulated Sensors". A pressure sensor utilizing this concept is shown in Figure 8-3.[9]

The reflective fiber optic sensor is described in detail in Chapter 3. It is a very precise, compact, noncontact displacement monitor. Therefore, the approach is ideal for monitoring diaphragm movement.[6] Figure 8-4 depicts a very simple pressure-sensitive diaphragm that can move relative to a reflective fiber optic probe. The response curve showing relative output

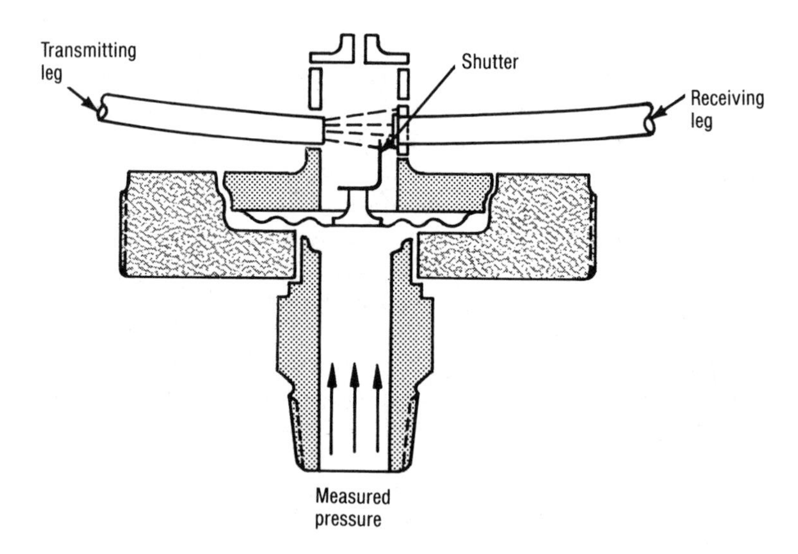

Figure 8-1
Transmissive Fiber Optic Pressure Sensor Using a Shutter to Modulate the Intensity

versus pressure is shown in Figure 8-5. The nonlinear response can be corrected in the electronics and/or by careful diaphragm (transducer) and fiber optic probe design. As an example, diaphragm-type pressure transducers are available with linear response behavior and little hysteresis. Such a transducer may typically have a diaphragm linear movement of 0.015 inch over a pressure range of 500 psi. In choosing a corresponding fiber optic reflective probe, the sensitivity and linearity are the primary design selection parameters. Referring to Figure 3-4 showing various reflective fiber optic sensor response curves, the front slope generally has better linearity and sensitivity than the back slope. To achieve high sensitivity and maximum linearity for the required displacement range, both the hemispherical and pair probes are candidates. Front slope linearity is generally maintained over about 70% of the range. Therefore, the hemispherical probe has a linear range of about 0.020 inch and a sensitivity of 0.04 arbritary intensity units per 0.001 inch of movement. The large fiber pair probe has a linear range of about 0.035 inch and a sensitivity of 0.023 arbritary intensity unit

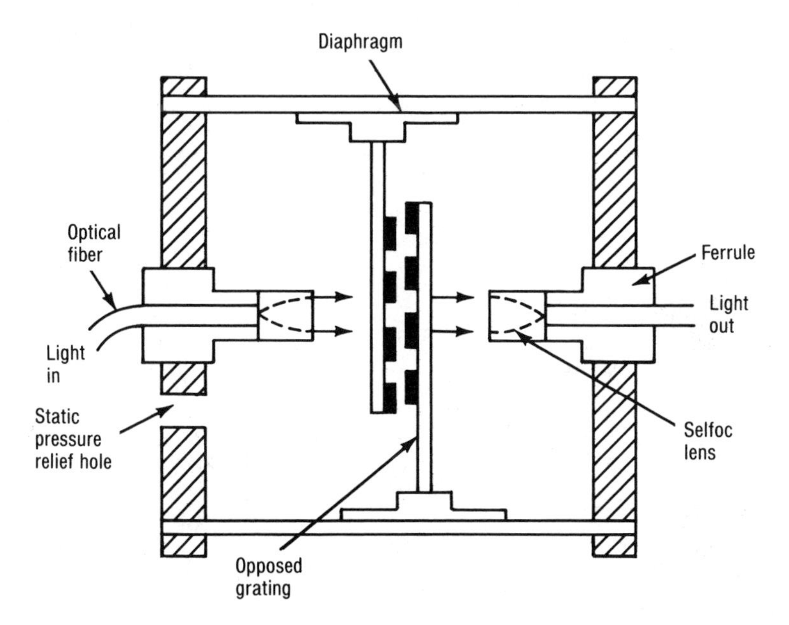

Figure 8-2
Transmissive Fiber Optic Pressure Sensor Using a Moving Grating to Modulate the Intensity

(© 1982 IEEE; reprinted with permission)

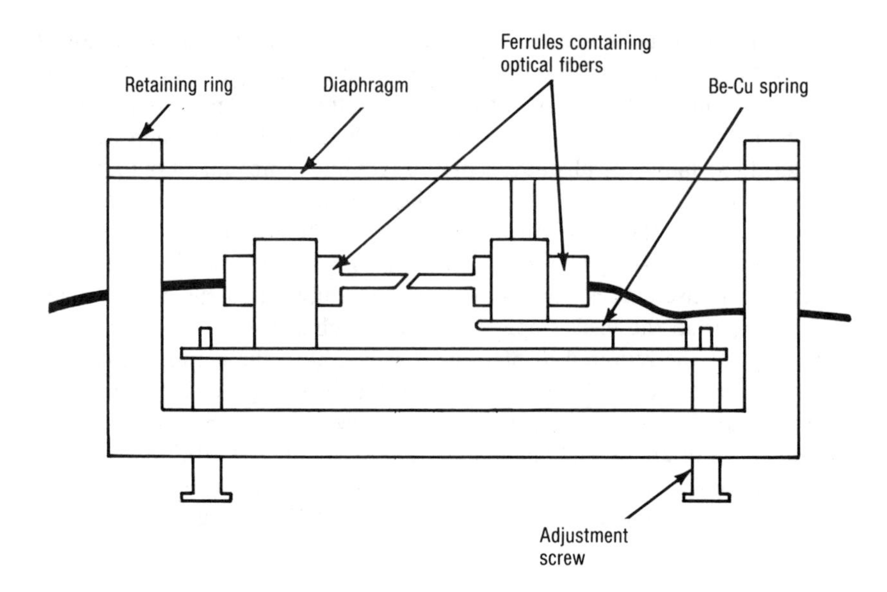

Figure 8-3
FTIR Pressure Sensor

(reprinted by permission from the Optical Society of America)

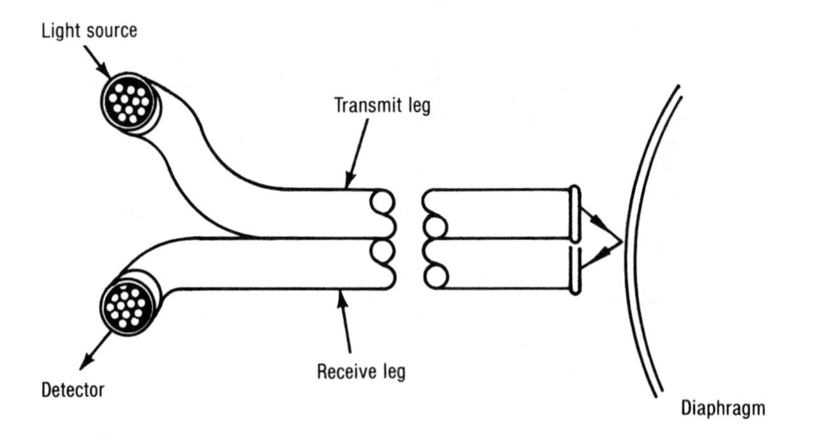

Figure 8-4
Reflective Fiber Optic Pressure Sensor Using a Diaphragm for Modulation

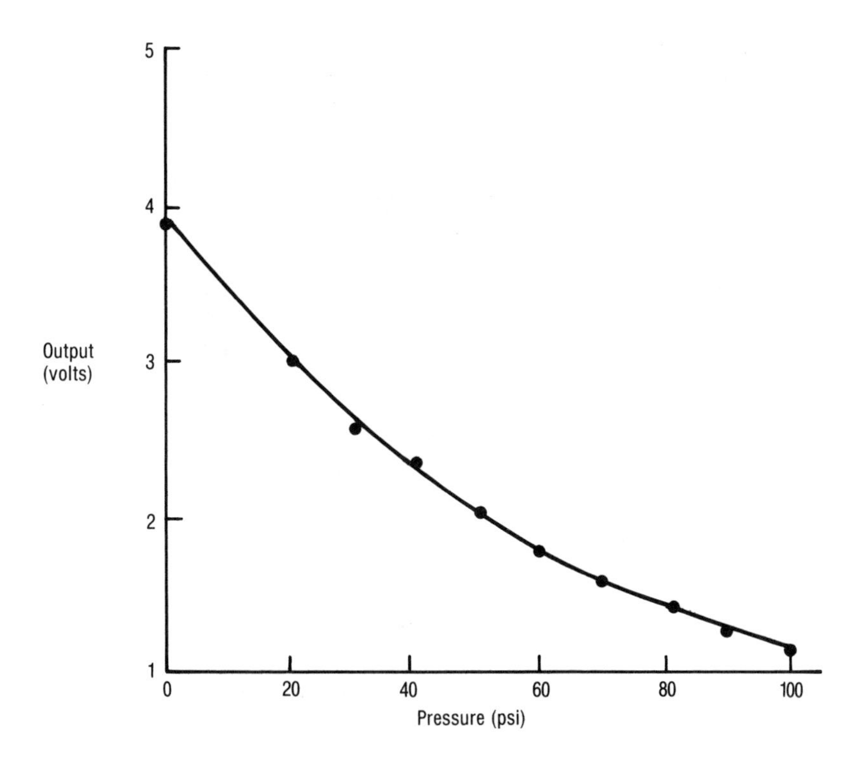

Figure 8-5
Response Curve for a Reflective Fiber Optic Sensor

per 0.001 inch movement. For this case, the hemispherical probe was selected. If a longer linearity range was required, the pair fiber probe would be used.

The noncontact nature of the reflective fiber optic sensor is especially important under dynamic conditions. Pahler and Roberts(7) have shown that, as the frequency of pressure probes increases up to the 100 Hz range, the fiber optic reflective sensor has an increasingly faster response than a bonded strain gage. The observation can be attributed to inertial effects of the diaphragm. The fiber optic unit has less mass and therefore faster response. This trend continues to much higher frequencies.

Another reflective sensor concept that has potential for pressure-sensing use is near total internal reflection, NTIR.(8) The configuration is shown in Figure 8-6. The end of the single-mode fiber is cut at an angle just below the critical angle. Light is reflected at the angled end and at the mirror surface, and the return light path (down the same fiber) is monitored. Pressure alters

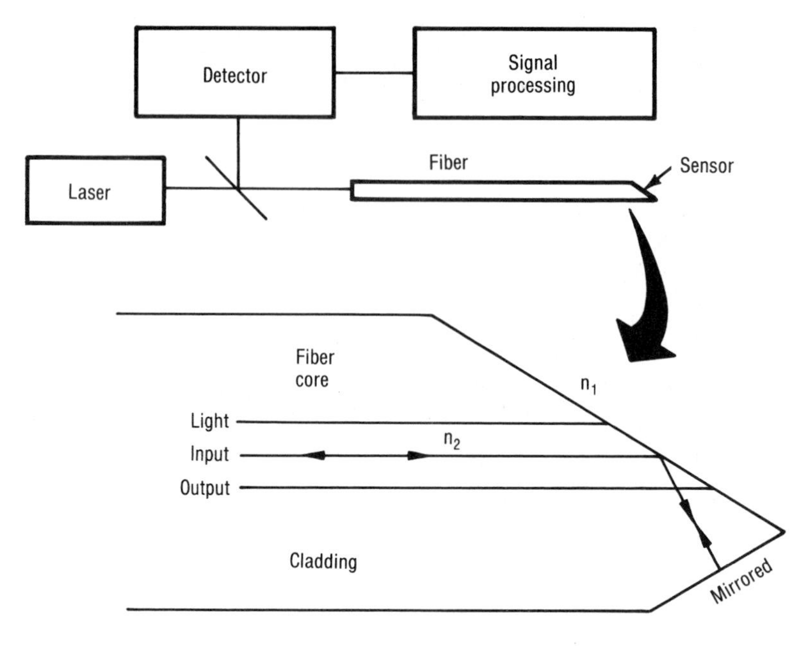

Figure 8-6
NTIR Fiber Optic Pressure Sensor

the refractive indices, n_1 and n_2 differently, resulting in a slight shift in critical angle; the resultant returned light intensity is modulated as a function of pressure. This sensor configuration has the distinct advantage of being extremely small.

MICROBENDING

The microbending concept is described in detail in Chapter 6 "Displacement Sensors." Since the concept is a displacement monitor, it is especially suited for pressure measurement. Figure 8-7 depicts a microbending pressure element.[8] High sensitivites for the sensor concept have been reported.[10,11]

The microbending approach can be used for the same pressure diaphragm specification as the reflective sensor previously described. Figure 6-21 shows a microbending fiber (5 point bending, 5.5 mm (0.22 inch) bend point spacing). The linear operation of the sensor occurs over approximately 0.015 inch. Since this value matches the diaphragm movement, the full diaphragm deflection would give a linear microbending modulation output. Generally, the linear displacement range of the sensor can be increased by increasing the bending point spacing.

A problem with microbending sensors is flow of the fiber coating under load, which causes relaxation of the fiber bending. This effect can be minimized with metalized coatings or by biasing the fiber by loading it prior to a displacement load. This bias load causes the coatings to flow; and since the load is never totally removed, the coating flow is not allowed to relax. Other disadvantages of the microbending approach compared to the reflective

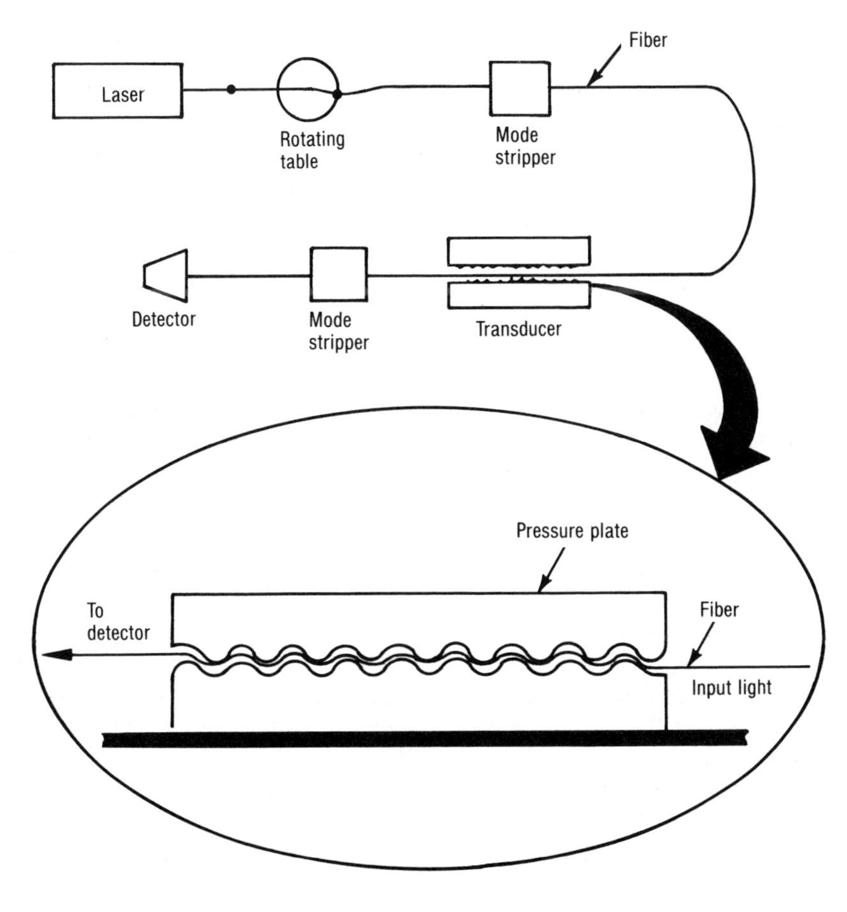

Figure 8-7
Fiber Optic Microbending Pressure Sensor

approach include the requirement to contact the diaphragm. This contact requirement limits dynamic response. Also, care must be taken that the lead fibers going to and from the transducer area are not sharply bent to prevent the leads from acting as sensors. This requirement prevents a highly compact sensor configuration. The major advantage is that microbending provides for a closed system, generally free from contamination problems.

INTRINSIC CONCEPTS

Generally, intrinsic fiber optic pressure sensors employ a pressure-dependent change in refractive index to affect either the fiber output aperture or the polarization-related intensity of the light. Figure 8-8 shows a polarization-pressure sensor.(12) The incoming laser light is polarized and transmitted through a short section of single-mode fiber mechanically loaded at an angle 45° to incoming polarized light. The fiber acts as a birefringent element, and the plane of polarization is rotated with the output intensity-modulated. The light intensity versus applied pressure is shown in Figure 8-9. The pressure level corresponds to a rotation of the plane of polarization, θ, with the intensity being proportion to $\cos^2\theta$ (see Figure 3-11). Figure 8-10 shows a similar approach with a birefringent element (nonfiber optic). Several birefringent materials with various stress optic coefficients can be used to obtain a given sensitivity or working range. The working range of the sensor is defined as the pressure range corresponding to the difference between a maximum and a minimum (½ period) of the intensity curve. The sensitivity is defined as the slope of the light intensity versus pressure curve. A birefringent element material with a high stress optic coefficient has increased sensitivity with decreased working range.

It is possible to construct a fiber in which, under pressure, the refractive index of the cladding increases relative to the core.(12, 13, 14) Under such conditions, the fiber has an increase of the critical angle at the core-cladding interface. Therefore, some of light, which would be carried when the fiber is not under pressure, is lost when pressure is applied and the output aperture is decreased. The change in output aperture can be detected and correlated with pressure.

A device using the refractive index change technique was fabricated with a plastic-clad silica fiber.(12) The pure silica core had a diameter of 50 microns and the silicone rubber cladding had an outside diameter of 200 microns. Ten loops of fiber were coiled around a 2 mm diameter rod and compressed between two flat plates. The fiber light was mode stripped (i.e., light traveling in the cladding was removed) before entering the transducer region so

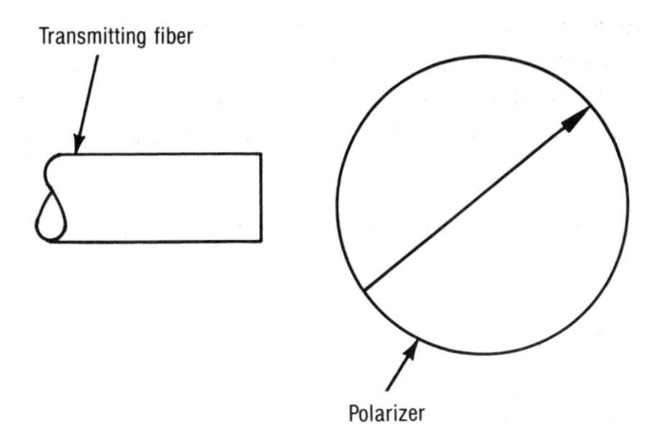

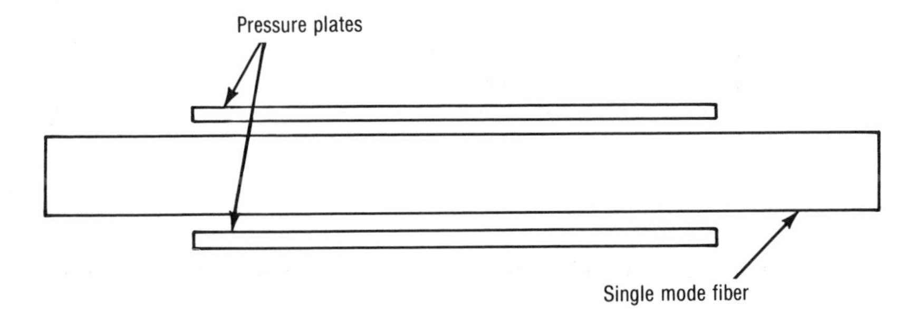

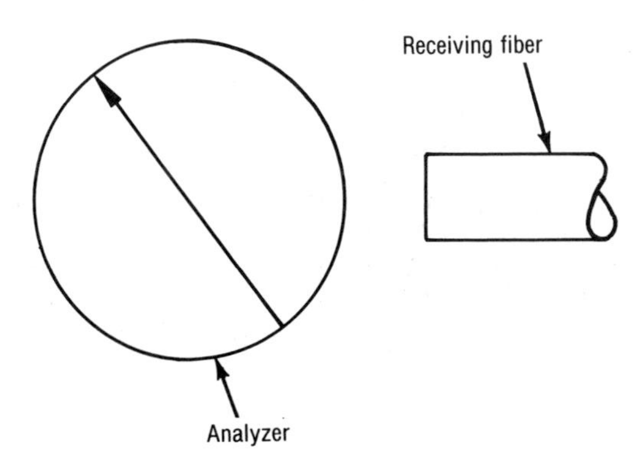

Figure 8-8
Intrinsic Polarizing Fiber Optic Pressure Sensor

that only pressure-coupled light would be detected. The output of the sensor versus applied pressure is shown in Figure 8-11. The refractive index difference between the silicone (1.41) and silica (1.46) is large. If a cladding with an index closer to silica is used, the sensitivity would be increased.

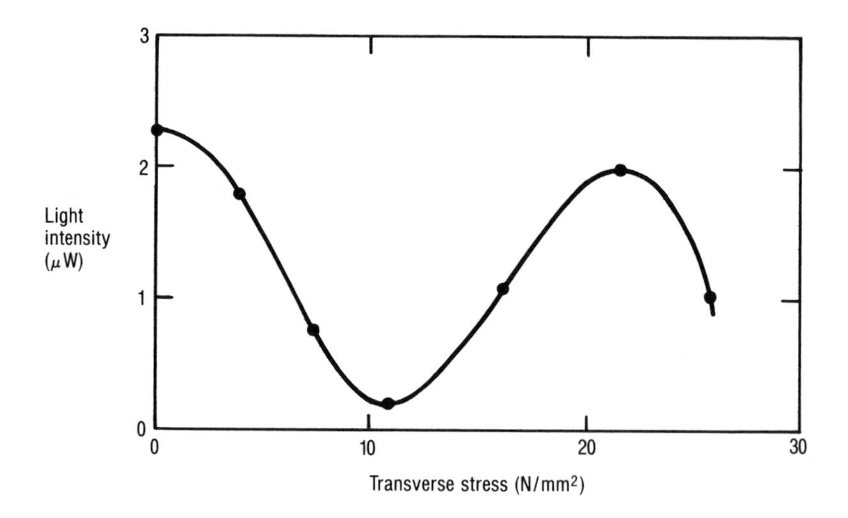

Figure 8-9
Light Intensity versus Applied Transduced Stress for the Intrinsic Polarization Sensor

(reprinted by permission from the Optical Society of America)

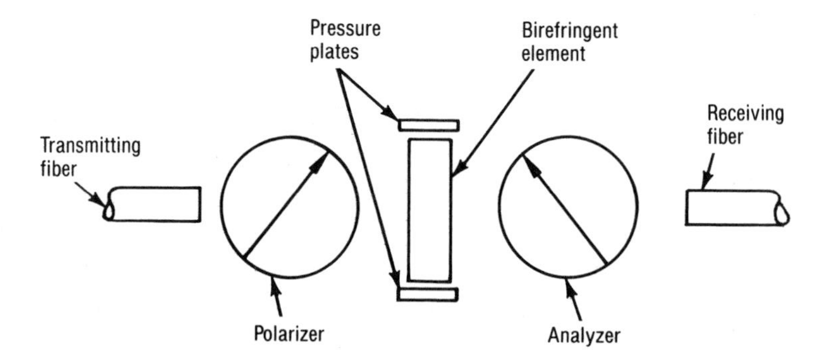

Figure 8-10
Polarizing Fiber Optic Pressure Sensor Using a Birefringent Crystal

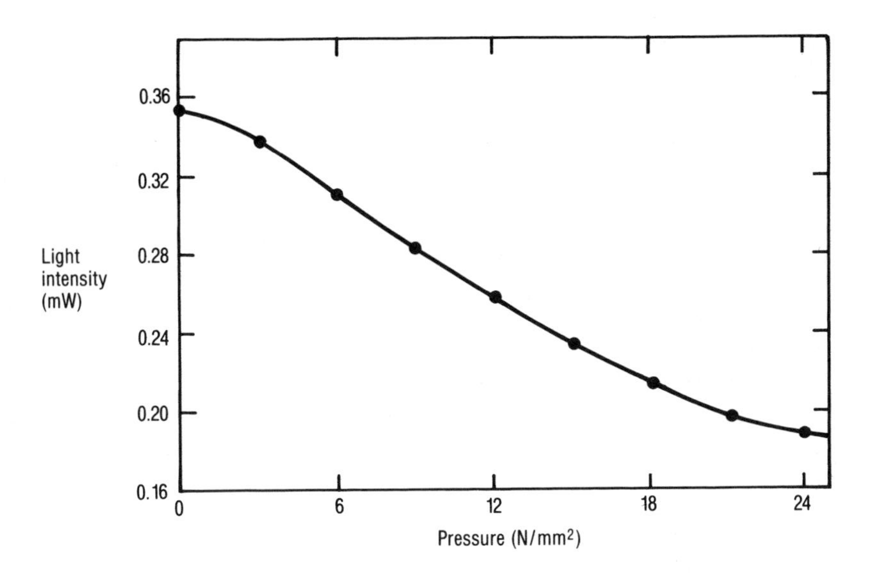

Figure 8-11
Light Intensity versus Applied Pressure for a Plastic Clad Silica Fiber
(reprinted by permission from the Optical Society of America)

INTERFEROMETRIC CONCEPTS

Most interferometric pressure sensors use the Mach-Zehnder configuration shown in Figure 8-12. The beam is split into a sensing leg and a reference leg. The reference leg is provided with modulation means. The sensing leg experiences the pressure that affects both fiber length and refractive index. The two beams are recombined and the phase modulation detected. The pressure sensitivity, Q, is defined as:(15)

$$Q = \frac{\Delta\phi}{\Delta P} \tag{8-1}$$

where ϕ is the phase angle, $\Delta\phi$ is the phase change, and ΔP is the pressure change. Since:

$$\Delta\phi = k(n\Delta L + L\Delta n) = kLn\left[\frac{\Delta L}{L} + \frac{\Delta n}{n}\right] \tag{8-2}$$

where k is the wave number, n is the core refractive index, Δn is the change in refractive index, L is the fiber length, and ΔL is the change in length. Substituting:

$$\phi = kLn \text{ (since } \phi = \frac{2\pi nL}{\lambda} \text{ and } k = \frac{2\pi}{\lambda})$$

in Equation 8-2 results in:

$$\Delta\phi = kLn \left[\frac{1}{L}\frac{\partial L}{\partial P} + \frac{1}{n}\frac{\partial n}{\partial P} \right] \Delta P \qquad (8\text{-}3)$$

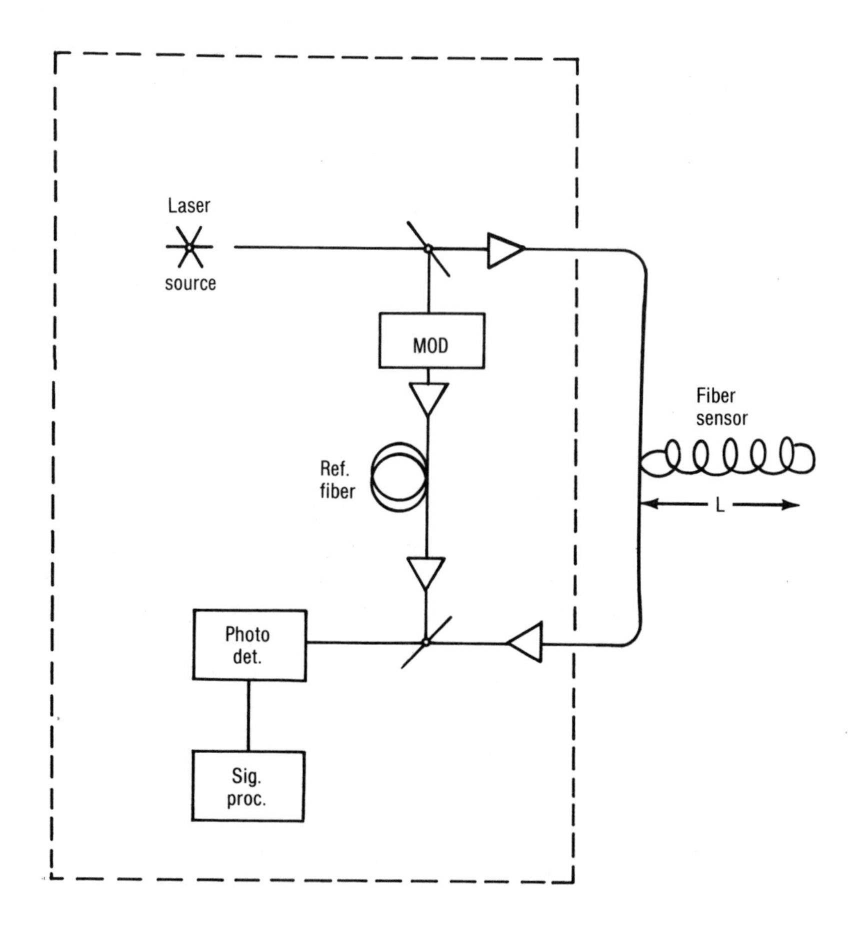

Figure 8-12
Mach-Zehnder Interferometer Used in Pressure Sensing

Therefore:

$$\frac{\Delta \phi}{\phi \Delta P} = \left(\frac{1}{L}\right)\frac{\partial L}{\partial P} + \left(\frac{1}{n}\right)\frac{\partial n}{\partial P} \tag{8-4}$$

For a given pressure change, it can be shown that a fiber core axial strain, ϵ_z and a radial strain, ϵ_r, develop.[5] As a result:

$$\frac{\Delta \phi}{\phi} = \epsilon_z - \frac{n^2}{2}\left[(P_{11} + P_{12})\,\epsilon_r + P_{12}\,\epsilon_z\right] \tag{8-5}$$

where P_{11} and P_{12} are the elasto-optic coefficients. The first term is related to the pressure sensitivity associated with length changes and the second term in brackets is related to the pressure sensitivity associated with the refractive index change. These effects are generally of opposite sign.

The various layers, both optical and protective, must be considered in determining the resulting strains in Equation 8-5. Therefore, the mechanical properties of the layers are directly related to the pressure sensitivity. Two examples are shown. First, Figure 8-13 shows the strains associated with a single-mode fiber coated with silicone rubber and hard plastic. The strain

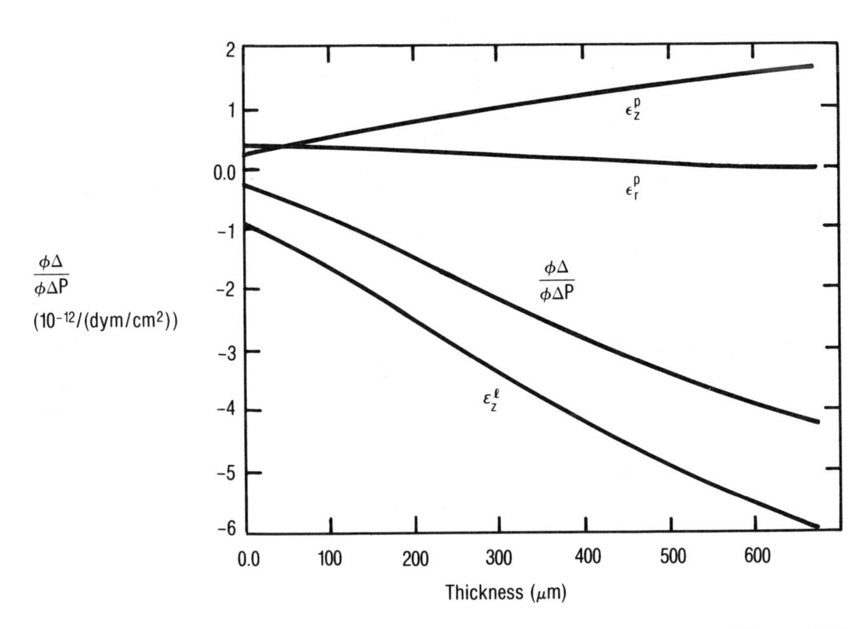

Figure 8-13
Pressure Sensitivity versus Coating Thickness (Plastic)

(© 1982 IEEE; reprinted with permission)

associated with length change, ϵ_z^l, gives the largest pressure sensitivity contribution. The strains that contribute to the refractive index change, ϵ_z^P and ϵ_r^P, give a smaller contribution of opposite polarity. As the plastic thickness increases, the pressure sensitivity increases rapidly with the dominant ϵ_z^l term. It was determined that the soft inner layer of silicone rubber had little effect. The hard outer layer is the prime contributor.

Figure 8-14 shows pressure sensitivity versus aluminum coating thickness.[16] The sensitivity changes in a manner opposite that of the plastic coating. As the aluminum coating becomes thick, the fiber becomes pressure-insensitive. The conclusion is that plastic coatings increase pressure sensitivity while metallic coatings reduce sensitivity.

As a general statement, coatings with high bulk modulus decrease pressure sensitivity. It has been shown[5] that as the coating becomes thick, the strain in the fiber due to hydrostatic pressure is due only to compressibility of the coating (inverse bulk modulus). Therefore, for thick coatings, the pressure sensitivity is a function only of bulk modulus and is independent of other elastic moduli. Figure 8-15 shows a plot of pressure sensitivity versus bulk modulus for fibers with typical coating thicknesses. High Young's modulus values in the coating result in a strong bulk modulus dependence.

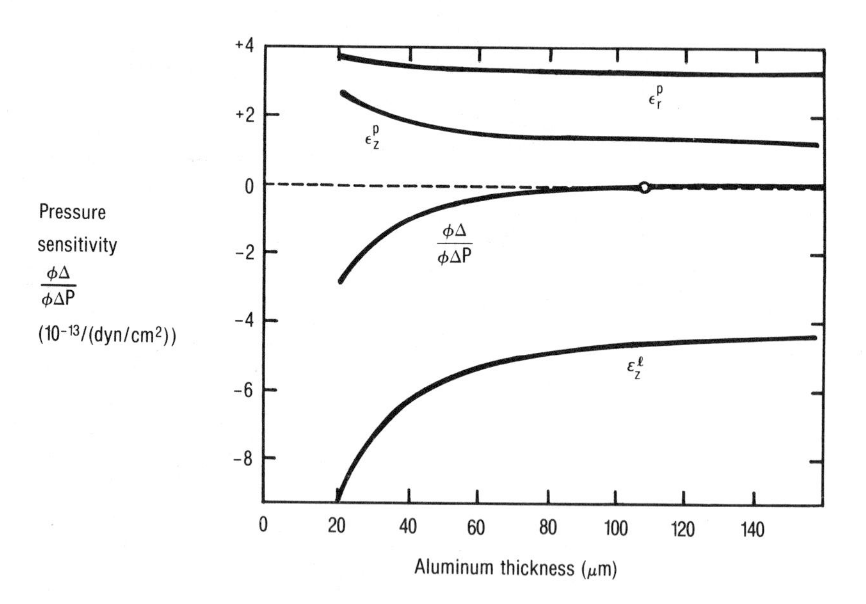

Figure 8-14
Calculated Pressure Sensitivity versus Coating Thickness (Aluminum)

(reprinted by permission from the Optical Society of America)

As Young's modulus decreases, so does the bulk modulus dependence. As a result, high pressure sensitivity requires coatings with high Young's modulus and low bulk modulus.

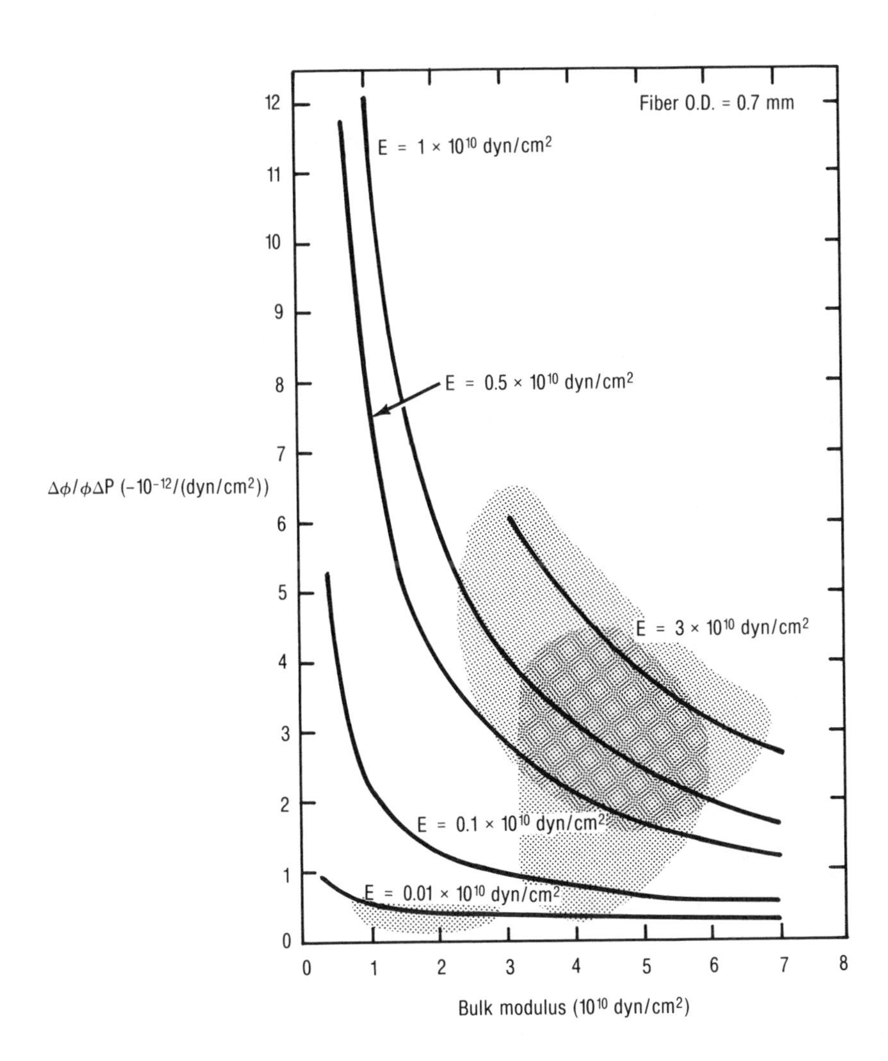

Figure 8-15
Calculated Pressure Sensitivity versus Bulk Modulus for Various Young's Moduli of the Outer Coating. Shaded areas: Plastics (Upper), UV Curable (Middle), Rubber (Lower).

Sensitivity is important in interferometric pressure sensors. The lead fibers and reference fibers must be desensitized to the pressure perturbations, while the sensing fiber in a localized area must be at maximum sensitivity for optimized sensor performance. This requirement can be accomplished by careful coating selection.

Another interferometric approach to pressure sensing uses the Fabry-Perot concept.[17] The system is shown schematically in Figure 8-16. Pressure on the diaphragm causes the distance between the diaphragm and the lens surface to change. The spacing change is measured with a Fabry-Perot interferometer. A mirror on the quartz plate surface and the quartz diaphragm serve as both ends of a Fabry-Perot cavity. The diaphragm movement with pressure causes a change in cavity length. The length change causes a shift in wavelength of the fringes in the reflected spectrum. The reflective light is focused onto the common fiber and transmitted to the optical decoder. The reflective light is directed by the beam splitter to the focusing grating, which acts as a single-element spectrometer. The spectrum is detected with the linear photodiode detector array. Figures 8-17(a) and (b) show an ideal and actual intensity versus wavelength curve for a Fabry-Perot interferometer. The minimum values correspond to fringes.

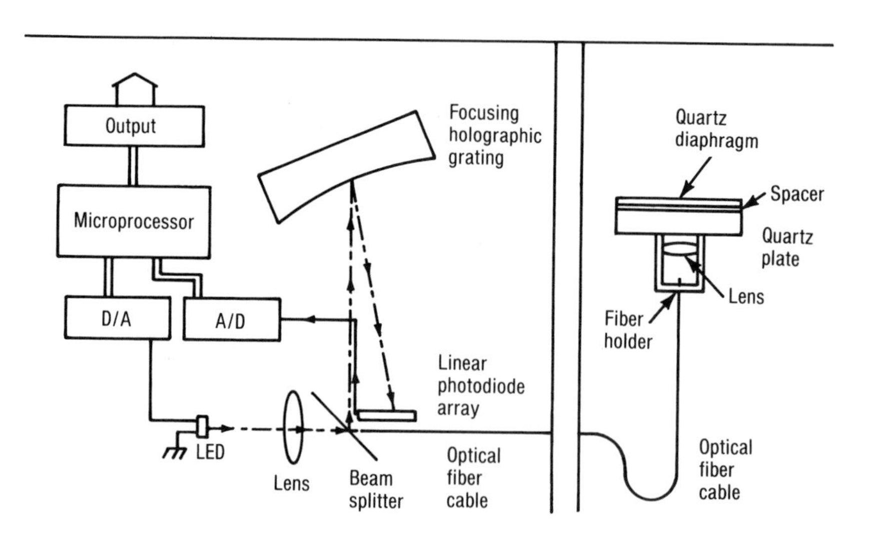

Figure 8-16
Schematic Diagram of Pressure Sensor Using Fabry-Perot Interferometer

When the cavity length changes, the intensity minimum (fringes) shifts to higher or lower wavelengths depending upon whether the cavity length is increased or decreased. The fringe pattern corresponds to a given pressure.

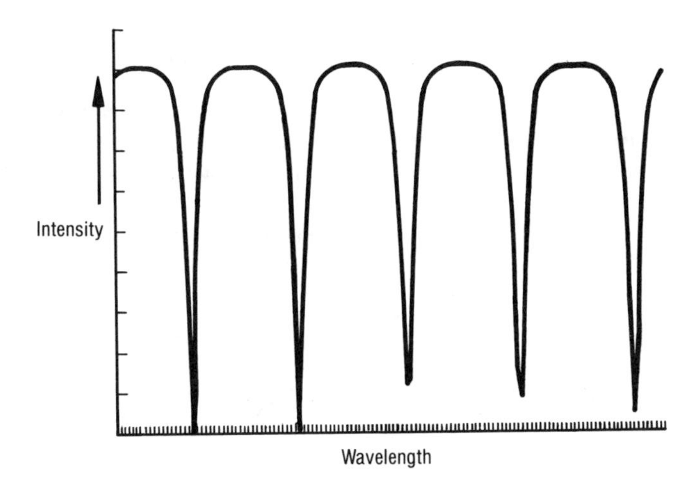

(a) Ideal

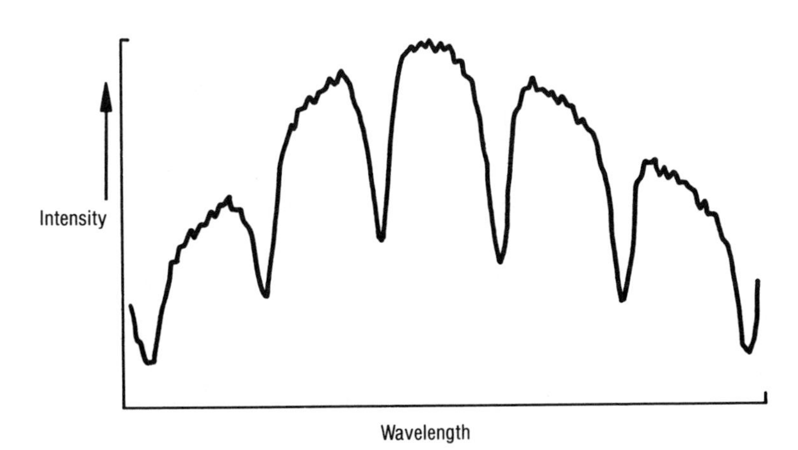

(b) Actual Pressure Sensor

Figure 8-17
Intensity Spectrum of Fabry-Perot Interferometer

This sensing scheme has several distinct advantages. The transducer head is made of quartz, which provides good temperature stability. Only one fiber is required, which allows the sensor to be compact. The sensor is insensitive to intensity variations in the lead fiber and can be made quite rugged.

APPLICATIONS

Much of the earlier work in fiber optic pressure sensors dealt with medical applications using the reflective probe in conjunction with a small diaphragm or a transmissive probe in conjunction with a shutter.[18,19,20,21,22]

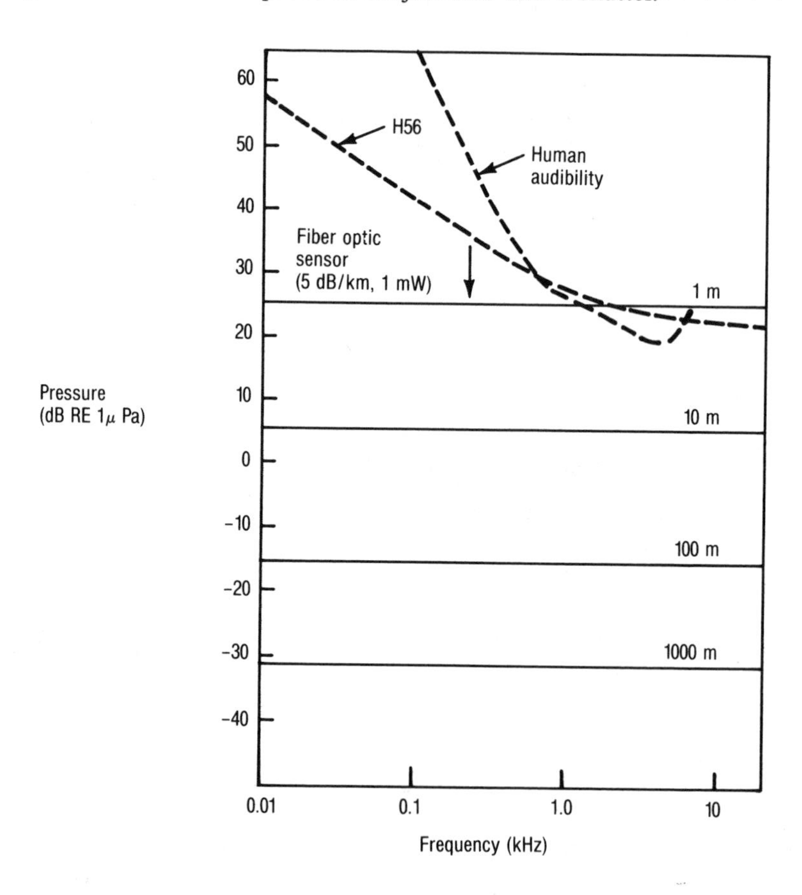

Figure 8-18
Minimum Detectable Pressure for Fiber Optic Acoustic Sensor

The flexibility and potentially very small sensor size makes the approach viable for *in vitro* blood pressure measurement. Pressure transducers have been built less than 1 mm diameter in size.

The immunity and inherent safety (explosion-proof) are the driving forces for application of fiber optic pressure sensors in industrial environments. Typical pressure ranges are 0–1000 psi, with accuracies between 1% and 0.1% including linearity and hysteresis. An important feature for industrial use is that the device may be remote up to hundreds of meters without electronic amplification. Therefore, specific applications include process monitoring in chemical and petrochemical plants.

Much of the funded research into fiber optic sensors has been directed at acoustic sensors for hydrophone applications. Figure 8-18 shows the minimum detectable pressure for an interferometric sensor.(5) The results are compared with a conventional piezoelectric (H56) hydrophone and the human ear. The interferometric sensor is independent of frequency variations and therefore can respond over a broad frequency range. Most diaphragm sensors have a much more narrow frequency band response.

REFERENCES

1. Gillum, D. R., 1982, *Industrial Pressure Measurements*, ISA.
2. Hall, J., 1978, "A Guide to Pressure Monitoring Devices," *Instrument and Control Systems*, pp 19–26.
3. Anonymous, "Non-Contacting Optical Sensor," *Bulletin HE/S-1*, Dresser Industries.
4. Porter, J. H., and Murray, D. B., 1972, "Fiber Optic Pressure Detector," U.S. Patent 3,686,958.
5. Giallorenzi, T. G., et al., 1982, "Optical Fiber Sensor Technology," *IEEE Journal of Quantum Electronics*, Vol. QE-18, No. 4, pp 626–665.
6. Krohn, D. A., 1983, "Fiber Optic Sensors in Industrial Applications — An Update," *Proceedings of the ISA*, Houston, TX, pp 877–890.
7. Pahler, R. H., and Robers, A. S., 1977, "Design of a Fiber Optic Pressure Transducer," *Transactions of the ASME*, pp 274–280.
8. Bucaro, J. A., and Cole, J. H., 1979, "Acousto-Optic Sensor Development," *Conference Record, Electronics and Aerospace Systems Conference*, Vol. 3, pp 572–600.

9. Spillman, W. B., and McMahon, D. H., 1980, "Frustrated-Total Internal Reflection Multimode Fiber Optic Hydrophone," *Applied Optics*, Vol. 19, No. 1, pp 113–117.

10. Fields, J. H., et al., 1979, "Multimode Optical Fiber Loss Modulation Acoustic Sensor," presented at the Optical Fiber Communications Meeting, Washington, DC.

11. Lagakos, N., 1981, "Intensity Sensors — Microbend," presented at the Fiber Optic Sensor System Workshop, Naval Research Laboratories.

12. Cielo, P., and Lapierre, J., 1982, "Fiber Optic Ultrasound Sensing for the Evaluation of Materials," *Applied Optics*, Vol. 21, No. 4, pp 572–575.

13. Stowe, D. W., and Christian, J. D., 1982, "Optical Transducer," U.S. Patent 4,354,735.

14. Stowe, D. W., and Christian, J. D., 1982, "Concentric Fiber Optic Transducer," U.S. Patent 4,363,533.

15. Davis, C. M., et al., 1982, *Fiber Optic Sensor Technology Handbook*, Dynamic Systems, Reston, Virginia.

16. Lagakos, N., Hickman, T. R., Cole, J. H., and Bucaro, J. A., 1981, "Optical Fibers with Reduced Pressure Sensitivity," *Optics Letters*, Vol. 6, No. 9, pp 443–445.

17. Belsley, K. L., Huber, D. R., and Goodman, J., 1986, "All Passive Interferometric Fiber-Optic Pressure Sensor," *Proceedings of the ISA*, Houston, TX, pp 1151–1158.

18. Frank, W. E., 1966, "Detection and Measurement Device Having a Small Flexible Fiber Transmission Line," U.S. Patent 3,273,447.

19. Strack, R. R., 1967, "Method of Fabricating a Pressure Transducer," U.S. Patent 3,503,116.

20. Polyanyi, M. L., 1963, "Devices for Measuring Blood Pressure," U.S. Patent 3,249,105.

21. Porter, J. H., and Murray, D. B., 1974, "Fiber Optic Pressure Detector," U.S. Patent 3,789,667.

22. Aagard, R. L., 1984, "Fiber Optic Pressure Sensor with Temperature Compensation and Reference," U.S. Patent 4,487,206.

Flow Sensors

9

INTRODUCTION

Flow measurement is a critical process control parameter in a wide range of applications such as engine control, power generation, and industrial processes. Often the environment is difficult. The sensor can be subjected to high electrical noise, explosive environments, relatively high temperature, and areas of difficult access. Fiber optic sensors have the ability to perform under these environmental conditions and to be the basis for several sensing approaches.

Four basic sensing concepts have been used for flow detection using fiber optics.[1] These concepts include the following:

(1) Rotational frequency monitoring of a paddle wheel or turbine in the flow field
(2) Differential pressure measurement across an orifice
(3) Frequency monitoring of a vortex-shedding device
(4) Laser Doppler velocimetry

These concepts will be discussed in terms of how they can be implemented with fiber optics.

TURBINE FLOWMETERS

Turbine flowmeters require that they be put in the flow path, which causes a flow obstruction. The meter also has moving parts. Both of these features are generally undesirable but the approach is accurate and repeatable and has broad industrial acceptance. Turbine flowmeters have a rotating device called a rotor, which is positioned in the flow stream. The rotor velocity is proportional to flow velocity of the fluid passing through the device.[2] A simple device is shown in Figure 9-1. The rotor rotational speed is monitored using a reflective fiber optic probe. As each vane of the rotor passes the probe, a reflected optical pulse is generated. Since the sensor is digital in nature, the problems associated with fiber optic intensity-modulated analog signals are eliminated.

DIFFERENTIAL PRESSURE FLOW SENSOR

For a differential pressure flow sensor, the volume of fluid is forced to flow through a restriction area with reduced cross section. The restriction causes an increase in flow rate at that point. The net effect is that there is a pressure drop associated with the restriction. The change in pressure is proportional to flow. A general description of differential pressure flow measurement can be found in References 2 and 3.

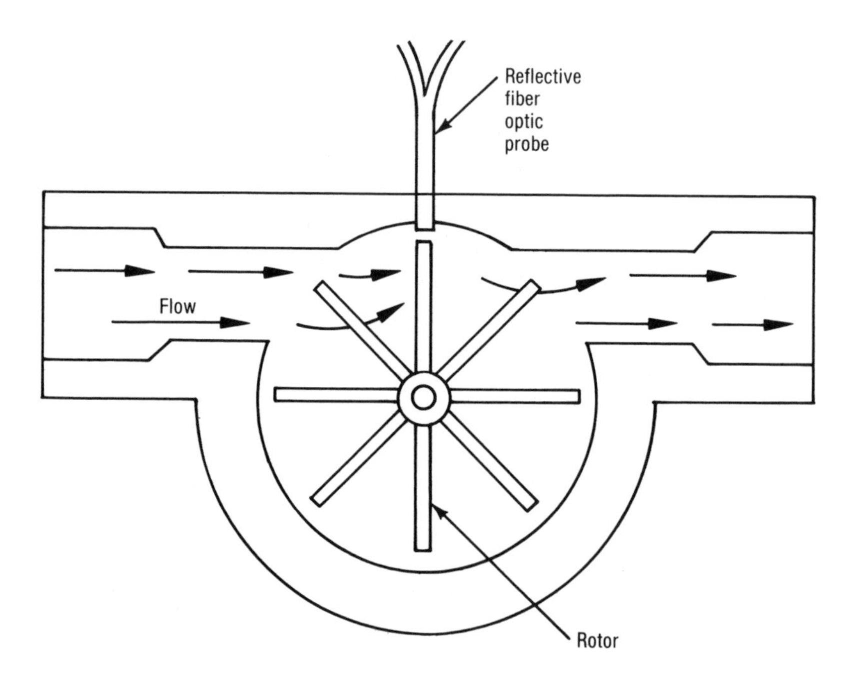

Figure 9-1
Turbine Flowmeter with a Reflective Fiber Optic Sensor

Figure 9-2 shows the concept of a differential pressure flow monitor using fiber optic pressure sensors. The pressure prior to the restriction is P_1; and the pressure after the restriction is P_2 with $P_1 > P_2$. As discussed in Chapter 8, several fiber optic pressure sensor concepts can be applied. In Figure 9-2, reflective sensors are employed.

The simplicity of differential pressure flow sensors gives them the potential for widespread use. They are especially well suited for low viscosity fluids and gases.

VORTEX-SHEDDING FLOW SENSOR

Figure 9-3 illustrates the concept of a vortex-shedding flow monitor. As a fluid passes over a bluff body, alternating vortices are generated from each side of the bluff body. The vortex formation causes a pressure pulse. The frequency of pressure pulses is proportional to the fluid velocity, as shown in the following equation:

$$f = SVd \tag{9-1}$$

where f is the pressure pulse frequency, S is the Strouhal number, a dimensionless constant that is flow rate dependent; V is the flow velocity; and d is the width of the bluff body. A detailed description of the approach is found in Reference 2. The unique feature of the sensor is that it is inherently

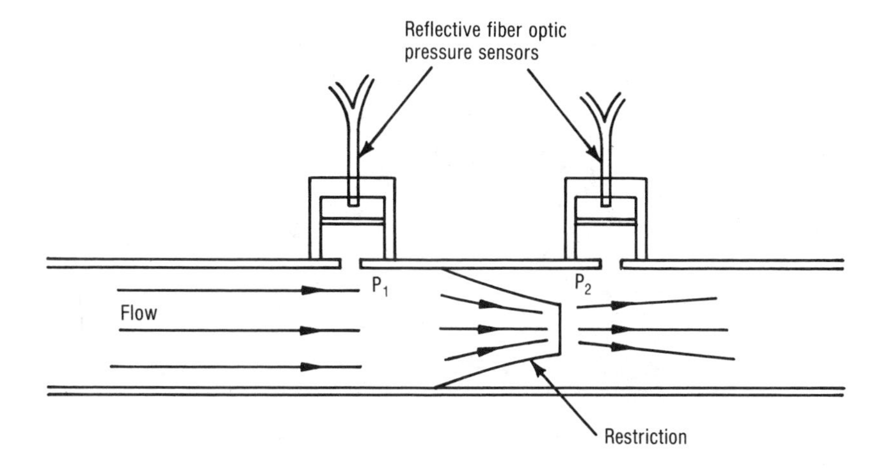

Figure 9-2
Differential Pressure Flow Sensor Using Reflective Fiber Optic Pressure Sensors

digital. Therefore, only a pressure excursion, not actually the magnitude of the pulse, needs to be determined. The position of the sensors is shown in Figure 9-4. Generally, access is limited; therefore, the reflective fiber optic technique for monitoring diaphragm movement is a logical choice for this application.

Another approach uses a microbending concept in a very simple configuration, as shown in Figure 9-5.[4] The optical fiber itself is the buff body. Chapter 6 discussed intensity modulation by means of microbending. This microbending approach uses the excitation pattern of light as the modulation mechanism. Changes in the pattern can be detected using a photodetector array; and the pressure perturbation frequency associated with vortex formation is determined.

Figure 9-6 shows a plot of shedding frequency (pressure pulse frequency) versus fluid flow rate (300 micron core fiber with N.A. = 0.37, water flow in 2.5 cm diameter pipe). The dashed line is the theoretically predicted curve. The measured flow is close to the theoretical case.

The distinct advantage of the approach is the very small size of the sensing element obstructing the flow path. A disadvantage is that it appears to be possible to have dead zones, which are associated with natural resonant frequencies in the sensor response.

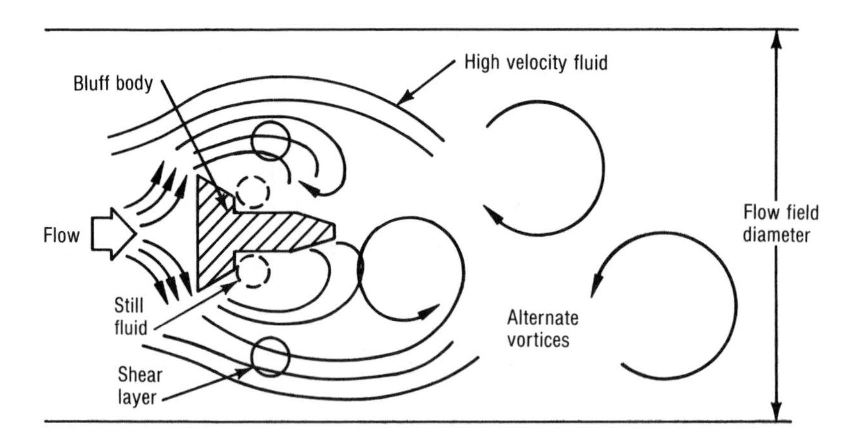

Figure 9-3
Shedding Vortex Concept

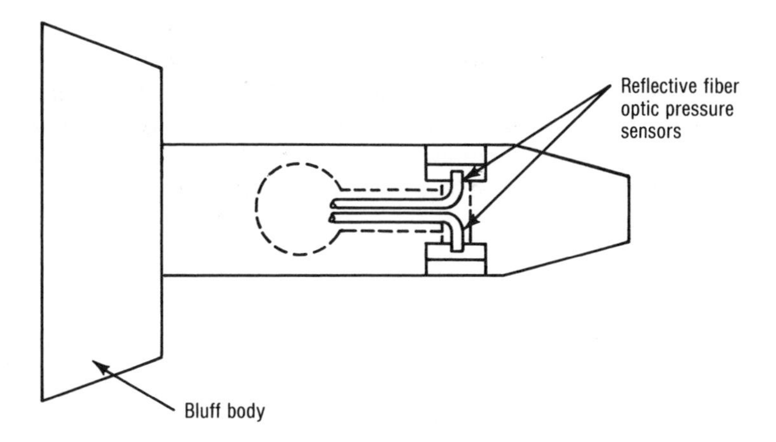

Figure 9-4
Reflection Fiber Optic Pressure Sensor in a Shedding Vortex Flow Monitor

LASER DOPPLER VELOCITY SENSORS

Laser Doppler velocimetry (LDV) provides for a noninvasive mechanism to measure fluid flow.(5) The dual beam method (see Figure 9-7(a)) provides two beams of equal intensity focused at a common point. The intersection of the two beams creates an interference with the fringes being generated. The fringe pattern is parallel to the bisector of the beams. As a particle passes through the fringe pattern (dark and light regions), light is scattered. The intensity across the laser beam associated with scattering is gaussian, as shown in Figure 9-7(b). In Figure 9-7(c), if the distance between fringes is d_f and the time to pass from one fringe to another is t, then the velocity component in a direction normal to the fringe is U_x. Therefore:

$$U_x - \frac{d_f}{t} = d_f F \tag{9-2}$$

where F is the frequency of fluctuation in scattered light intensity, referred to as the Doppler frequency.

The arrangement in Figure 9-7(a) requires constant polarization of the input light. Therefore, a polarization-maintaining fiber (discussed in Chapter 4 — Phase Modulated Sensors) is on the input side. A large core, multimode, high N.A. fiber is used for maximum collection efficiency.

Fiber optic-based LDV systems have distinct advantages. The use of fiber optics eliminates the need for bulkier optical components and allows for a very compact sensor. The technique is noninvasive. The approach measures velocity directly. It does not use flow properties in the measurement and, therefore, does not require calibration. The concept also has a very broad dynamic range, from low speed biological flows to supersonic levels.

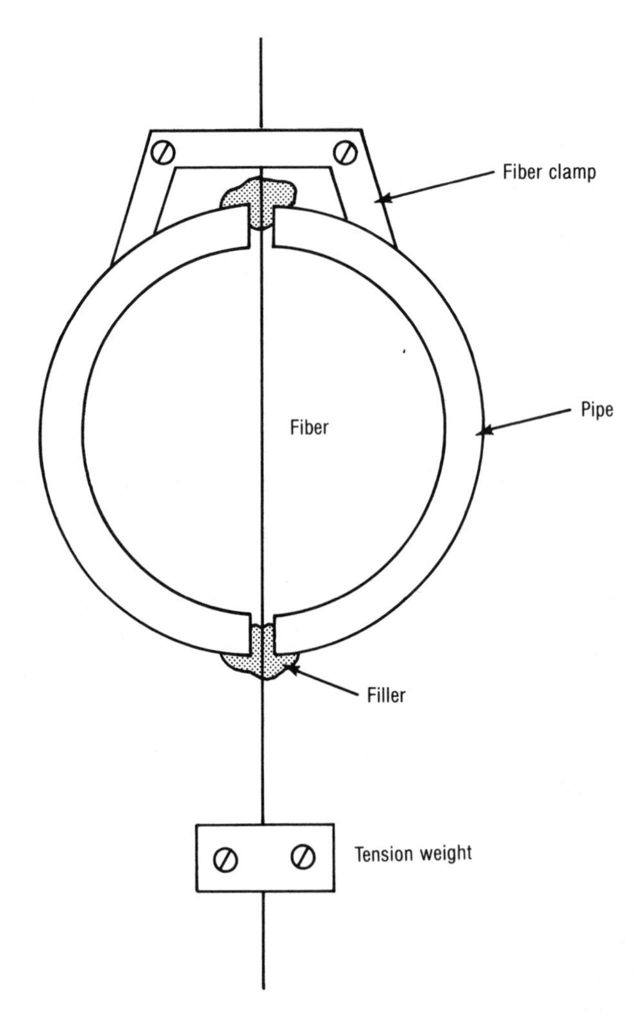

Figure 9-5
Shedding Vortex Monitor Using a Microbending Sensor
(© 1981 IEEE; reprinted with permission)

APPLICATIONS

Generally, the flow sensor concepts described have potential use primarily for industrial and power utility applications. Engine control applications are in an earlier product development stage, due in part to this application being more cost sensitive.

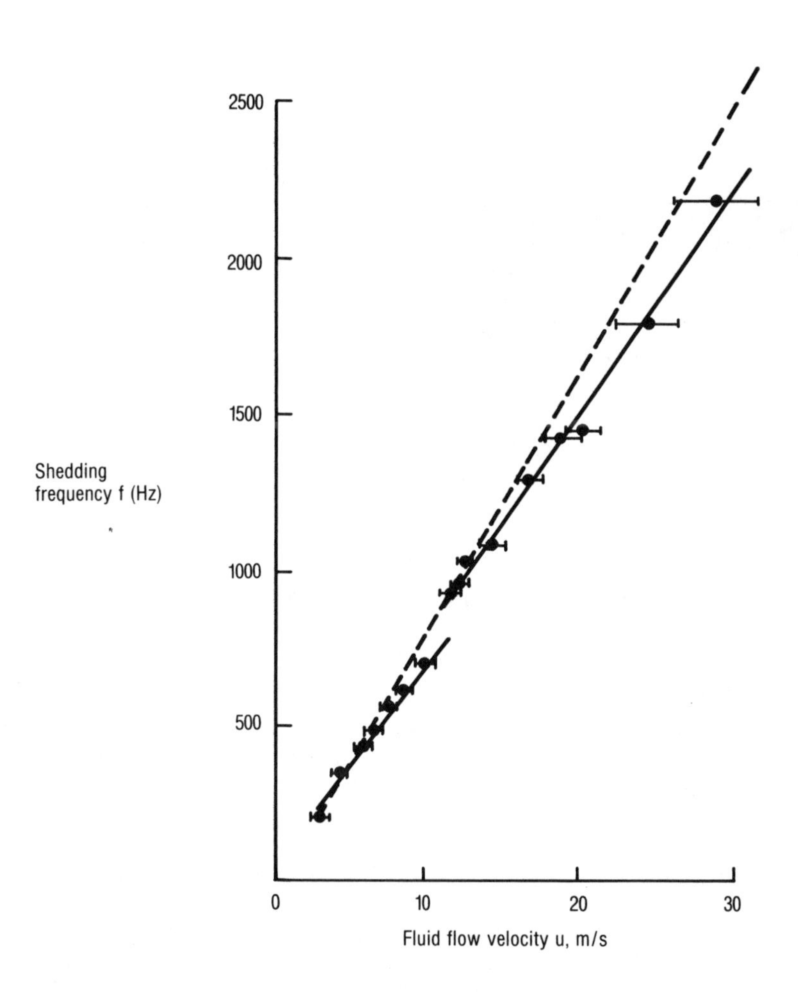

Figure 9-6
Shedding Frequency versus Fluid Flow Rate

(© 1981 IEEE; reprinted with permission)

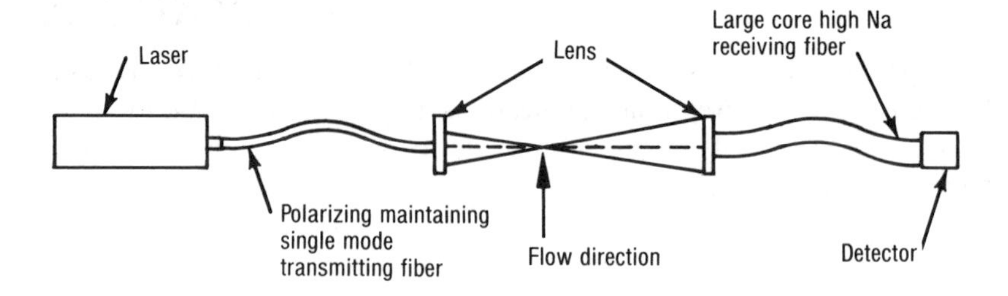

(a) Sensor Configuration (Dual Beam)

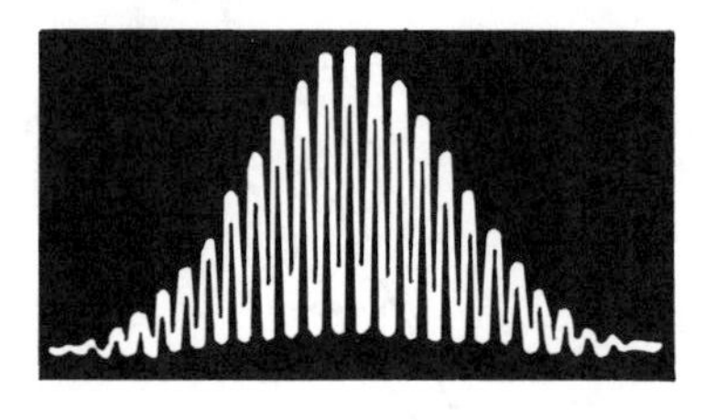

(b) Light Intensity Across Laser Beam

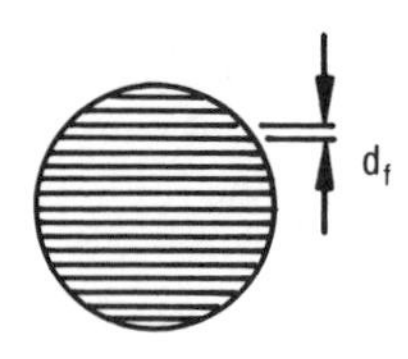

(c) Fringe Pattern at Beam Crossing Point

Figure 9-7
Principle of Operation of an LDV System

The immunity from electrical and environmental noise as well as the inherent safety of fiber optic sensors will be the driving force for their use.

The laser Doppler flow sensor has a much broader application range. It has been used for monitoring blood flow, biological cell flow, flow associated with industrial processes such as paper and steel manufacture, and flows associated with ultra high speed processes such as are found in aircraft applications.

REFERENCES

1. Giallorenzi, T. G., Buccaro, J. A., Dandridge, A., and Cole, J. H., 1986, "Optical-Fibers Sensors Challenge the Competition," *IEEE Spectrum*, pp 45–49.

2. Spitzer, D. W., 1984, *Industrial Flow Measurement*, ISA, Research Triangle Park, N.C., pp 209–219.

3. Gillum, D. R., 1982, *Industrial Pressure Measurement*, ISA, Research Triangle Park, N.C., pp 54–62.

4. Lyle, J. H., and Pitt, C. W., 1981, "Vortex Shedding Fluid Flowmeter Using Optical Fiber Sensor," *Electronics Letters*, Vol. 17, No. 6, pp 244–245.

5. Menon, R. K., 1986, "Velocity Measurement Using Fiber Optics," *Proceedings of the ISA*, Houston, TX, pp 1185–1195.

Level Sensors

10

INTRODUCTION

Level sensors are categorized as switches for high/low level and leak detection or as magnitude sensors for actual liquid level. Fiber optic devices have found more applications in the former category.

Liquid level is one of the prime process control parameters, especially in the petrochemical and chemical industries. The explosive nature of many of the processes makes fiber optics especially desirable. Leak detection and fuel level are required in many military applications. The immunity to EMI and RFI is the driving force for fiber optic usage in these applications.

Fiber optic sensors that use light interaction with the media for which the level is being measured work best in relatively clean and clear liquids. Dirty or somewhat opaque liquids such as crude oil and paints tend to foul the optics and blind the sensor, as do solids in powder form. Level measurement for these types of materials is best achieved by sensors that are used in pressure-related level measurements.

The concepts that can be used in conjunction with fiber optic level sensors include:[1,2]

(1) Sight glasses
(2) Force (buoyancy)
(3) Pressure (hydrostatic head)
(4) Reflective surface
(5) Refractive index change

SIGHT GLASS LIQUID LEVEL SENSOR

Boilers generally use a bi-color visual sight gage for water level.[3] A prism is used in such a way that if steam is in the gage port, red light is observed. Water in the port causes the red light to be refracted, with only green light transmitted. Such a device is shown in Figure 10-1. The location of the sensing device is typically at the boiler site, which is normally inconvenient for central monitoring. If the sensor is equipped with large core, large N.A. fibers, then the green/red water/steam signal can be transmitted to a central monitoring area, as shown in Figure 10-2. The interface of the fibers and color gage ports occurs within a hood to minimize any ambient light interference. The sensor is passive, with no moving or electrical components other than the illuminator. This approach is required by many current safety codes. Since the level of transmitted light must be high enough to be visible, the remote distance is limited to about 100 meters.

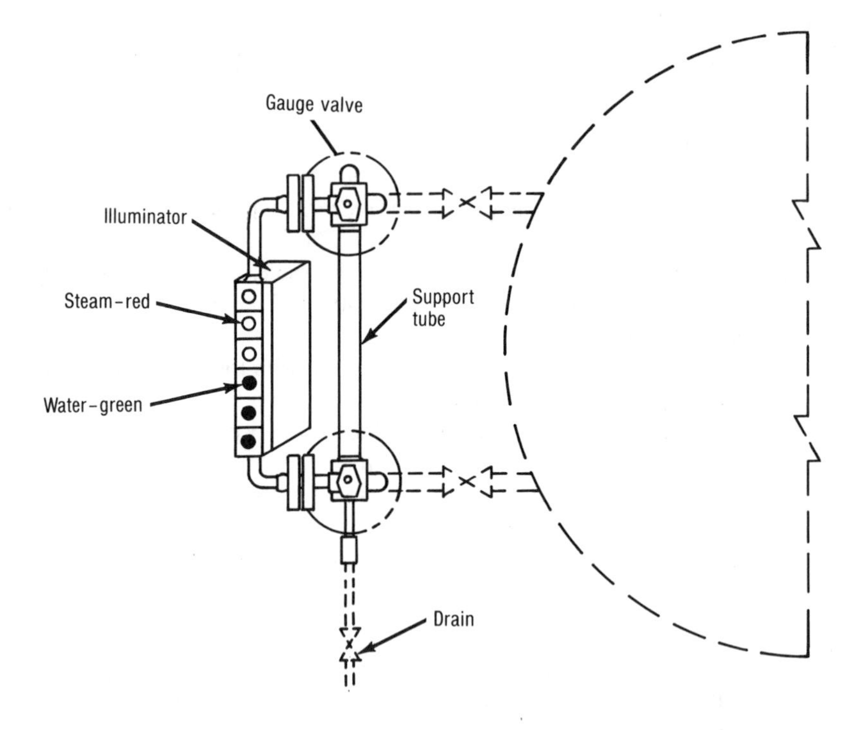

Figure 10-1
Bi-color Liquid Level Gauge

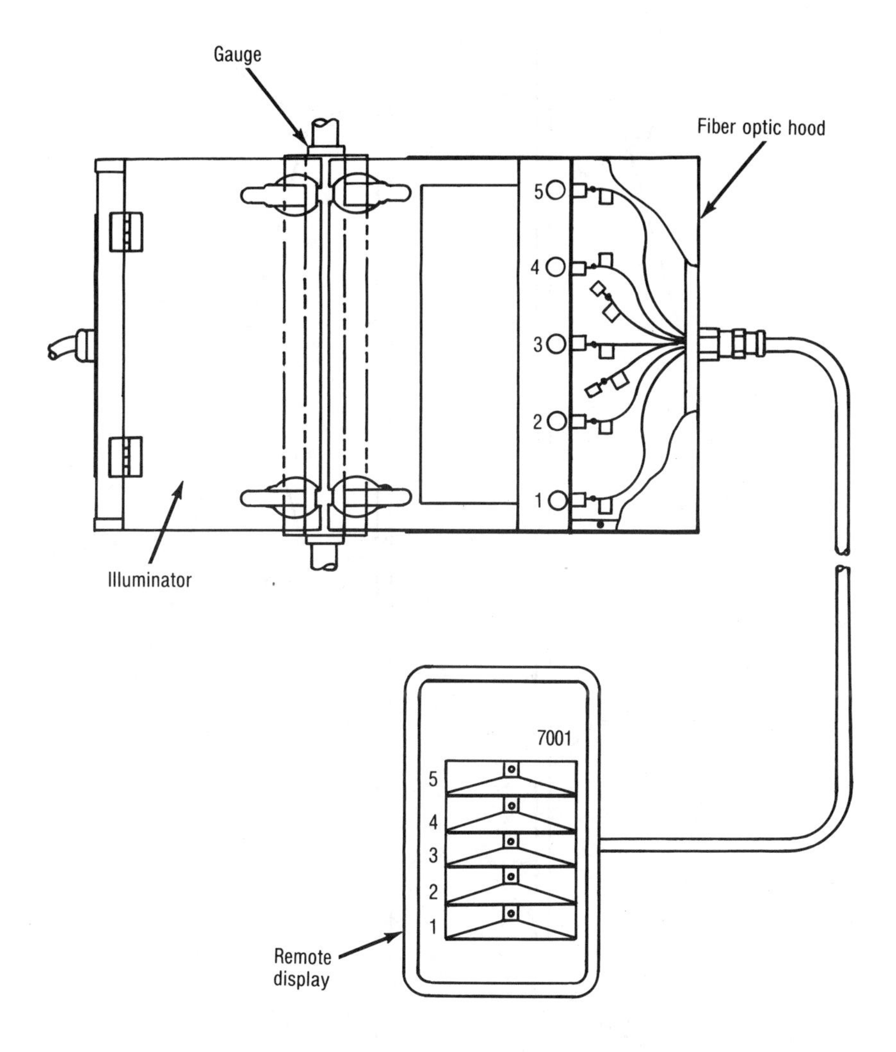

Figure 10-2
Fiber Optic Water Level Gauge Viewing System

Another simple approach for a sight glass uses a transmissive system, shown in Figure 10-3. As the liquid interrupts the light path, the level is determined.

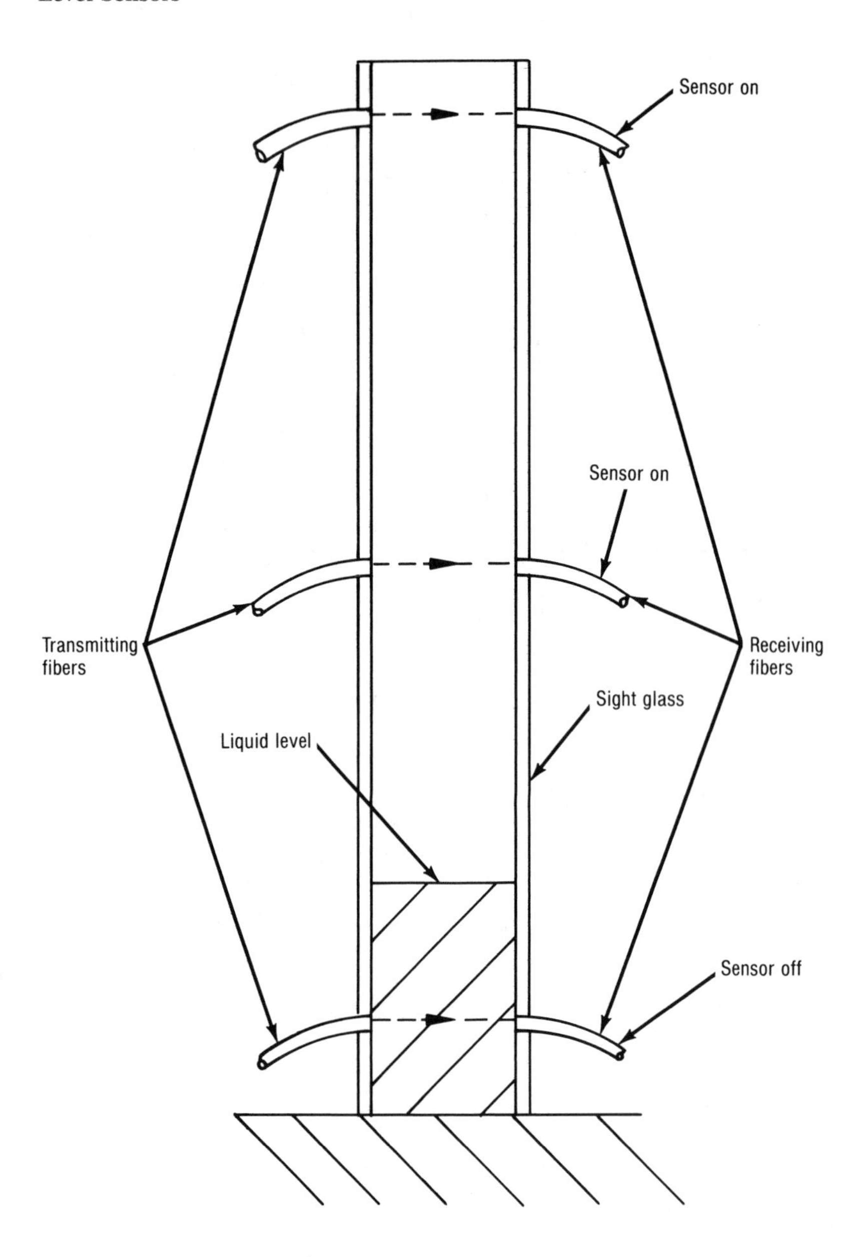

Figure 10-3
Transmissive Liquid Level Sensor for a Sight Glass

FORCE (BUOYANCY)

The force or buoyancy-type liquid level sensor often uses a float to track the liquid level. The simple float device shown in Figure 10-4(a) is designed to activate a reflective fiber optic sensor as a reflective target attached to the float passes by the probe. Figure 10-4(b) shows a similar device using a transmissive fiber optic sensor.[4] Sensors of this type are generally used for high or low level detection only. However, if the interruption device is made in the form of a transmissive binary code plate and multiple transmission-sensing fibers are employed, the actual height of the liquid can be determined in a digital format (see Figure 10-5). For an eight-bit code, the level can be resolved to one part in 256.

The actual height can be determined using a reflective sensor in an analog mode. Figure 10-6 shows the sensor position. As discussed in Chapter 6, a lensed reflective probe can provide an analog signal over a 5-inch span. Therefore, unless mechanical linkage is used, the maximum range for a sensor of this type is 5 inches.

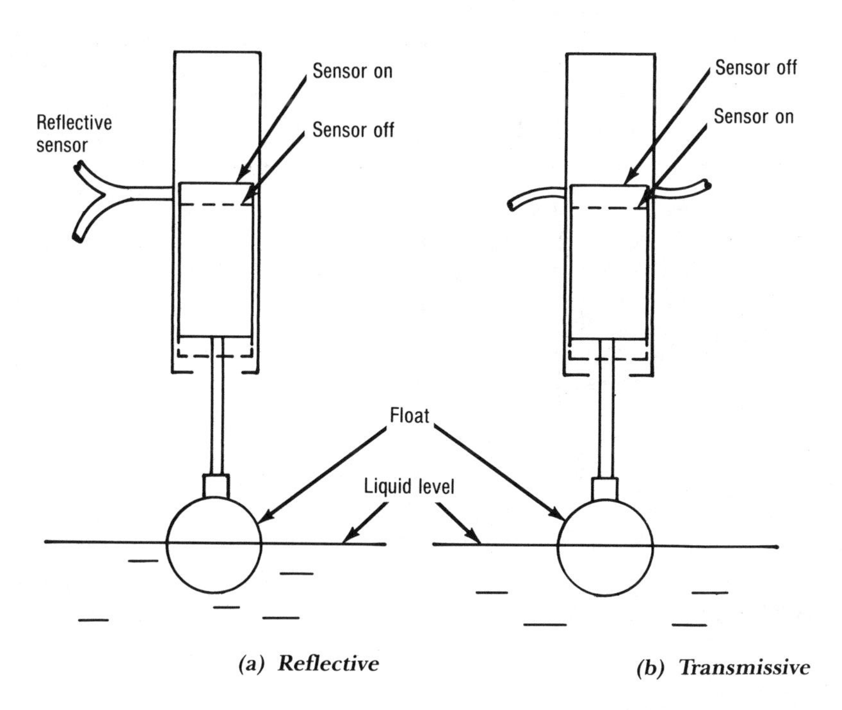

(a) Reflective *(b) Transmissive*

Figure 10-4
Liquid Level Switch

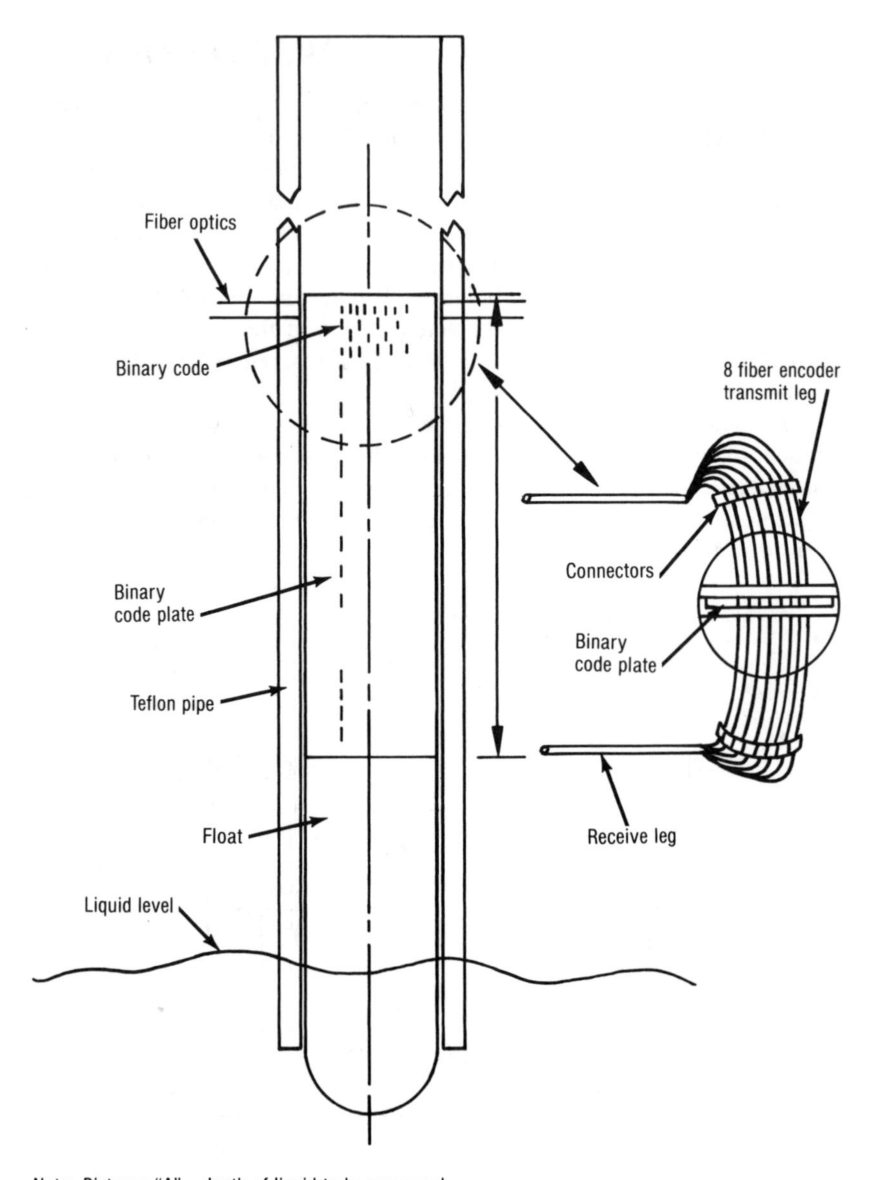

Note: Distance "A" = depth of liquid to be measured

Figure 10-5
Fiber Optic Liquid Level Sensor (Digital)

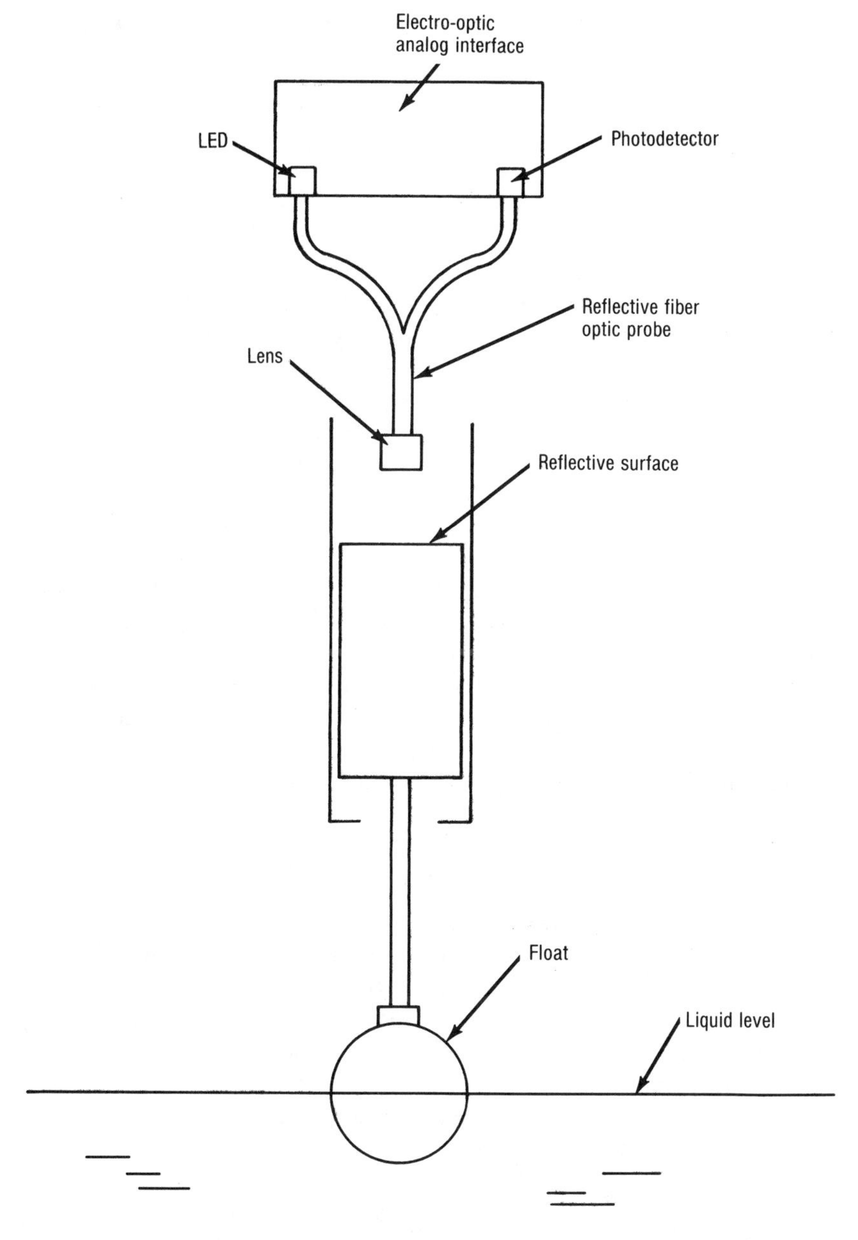

Figure 10-6
Liquid Level Sensor Using a Reflective Fiber Optic Probe in an Analog Mode

PRESSURE (HYDROSTATIC HEAD)

Liquid level can be determined by measuring the hydrostatic head: the pressure exerted by the level of liquid. The equation relating level height, H, and pressure, P, is given by:

$$H = \frac{P}{\rho SG} \tag{10-1}$$

where SG is the specific gravity of the liquid and ρ is the density of water. As examples,[2] a liquid level of 30 feet of water (SG = 1) equals 13.0 psi; 30 feet of oil (SG = 0.65) equals 8.44 psi, and 30 feet of brine solution (SG = 1.3) equals 16.9 psi. Due to higher pressure levels, high density liquids have a higher degree of accuracy in level measurement than low density liquids.

A reflective fiber optic sensor used with a diaphragm has been used in liquid level measurements. Figure 10-7 shows the sensor schematically. Typical water tank levels range from 20 to 60 feet, a pressure range of 0 to 26 psi. Resolution to 6 inches corresponds to a sensor accuracy of ± 0.21 psi ($\pm 1\%$).

The measurement is complicated if the tank is pressurized. The system requires two pressure sensors: one at the top of the tank to measure pressurization and one at the bottom to measure hydrostatic head plus pressurization, as shown in Figure 10-8. The difference in pressure values corresponds to the liquid level. A difficulty encountered is the fact that if the pressurization is high, small errors in pressure measurement cause large errors in liquid level when the difference is considered.

SURFACE REFLECTANCE

Surface reflectance techniques use the reflectance of light off the surface to determine liquid level.[5] For an incident beam of light striking the surface, the position of the detected beam is a function of liquid level, as shown in Figure 10-9. It is relatively straightforward to use optical fibers to remotely locate the electronic components (light source and photodetector) away from the measurement area.

A large dynamic range or very high sensitivity would be costly due to the large number of detection elements required. However, the approach is especially attractive for corrosive liquids or high temperature liquids where contact is not possible. As an example, such an approach can be used in a tank to measure the level of molten glass. If the light source is modulated (pulsed), preferably in the infrared only, the pulsed signal and not the ambient light from the molten glass, is detected. Molten glass also tends to

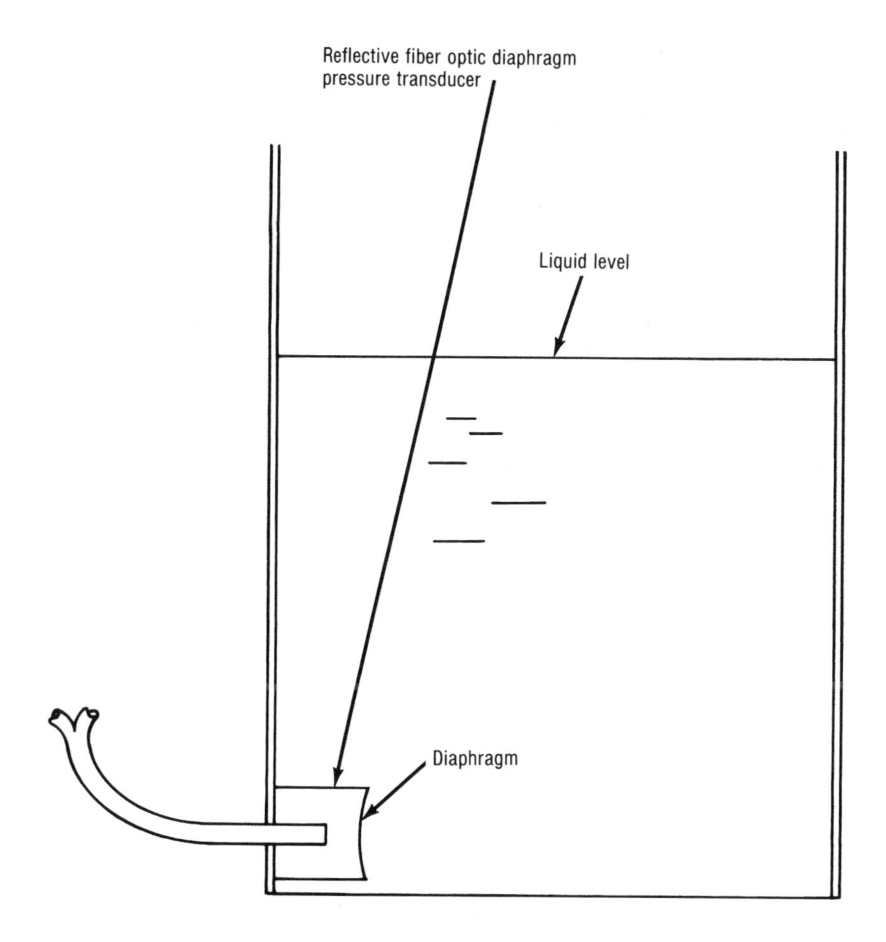

Figure 10-7
Liquid Level Sensor Using a Fiber Optic Pressure Transducer

be highly viscous with a smooth surface, making surface reflectance effective. Low viscosity liquids subject to surface vibration and ripple are not candidates for this measurement approach.

REFRACTIVE INDEX CHANGE

Refractive index change liquid level sensors function by transmitting light to a prism, typically quartz (refractive index = 1.46). In the medium of air (refractive index = 1.0), the prism acts as a fiber optic with the air being the cladding. The prism tip is shaped to promote back reflection, which is detected. In the presence of a liquid, the light is not totally internally reflected but is passed into the liquid.

Figure 10-10 shows the concept graphically.[6] Figure 10-11 shows the prism used, with the transmission and receiving legs being low loss optical fibers that allow remotability.[7]

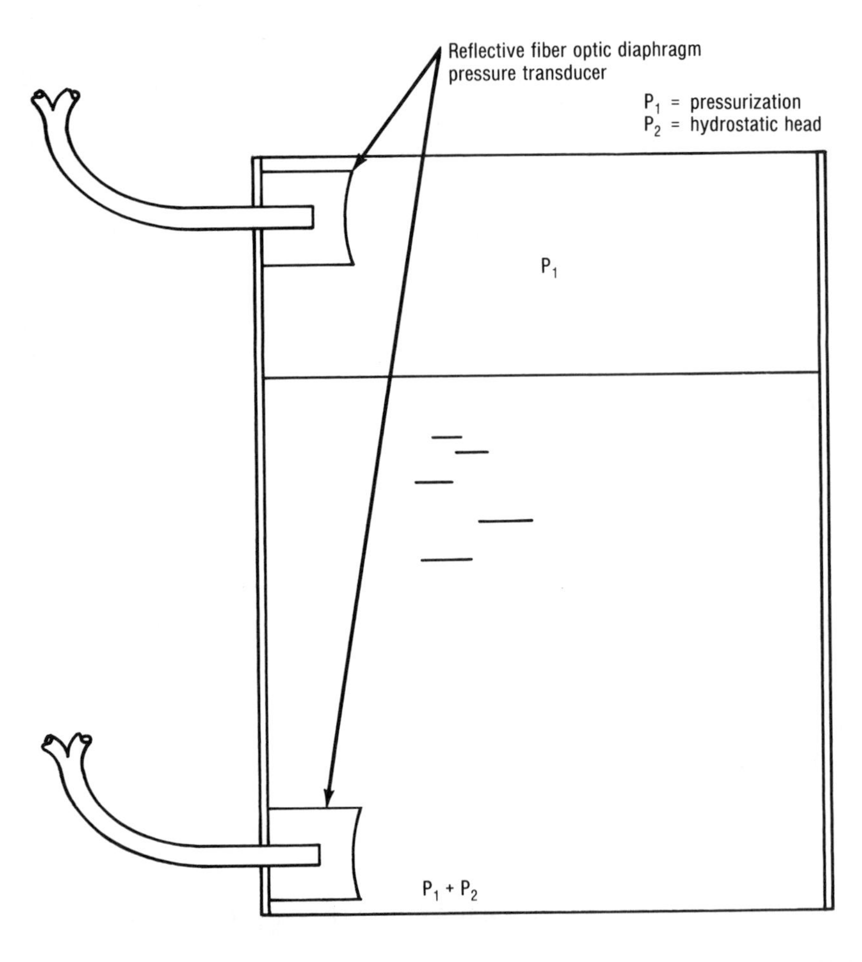

Figure 10-8
Liquid Level Sensor Using Fiber Optic Pressure Transducers at the Top and Bottom of a Pressurized Tank

Sensors of this type are point sensors. They are ideal for high/low monitoring as well as leak detection. A very small amount of liquid is required for switching. To achieve multilevel sensing, several sensors of various lengths can be ganged together.

The sensors described above are used in a digital mode to determine the absence or presence of liquid. However, since the sensor is analog in nature and provides an output that is a function of the refractive index of the liquid, the sensor can descriminate between liquids. The outputs for various liquids and air are given in the Table 10-1.

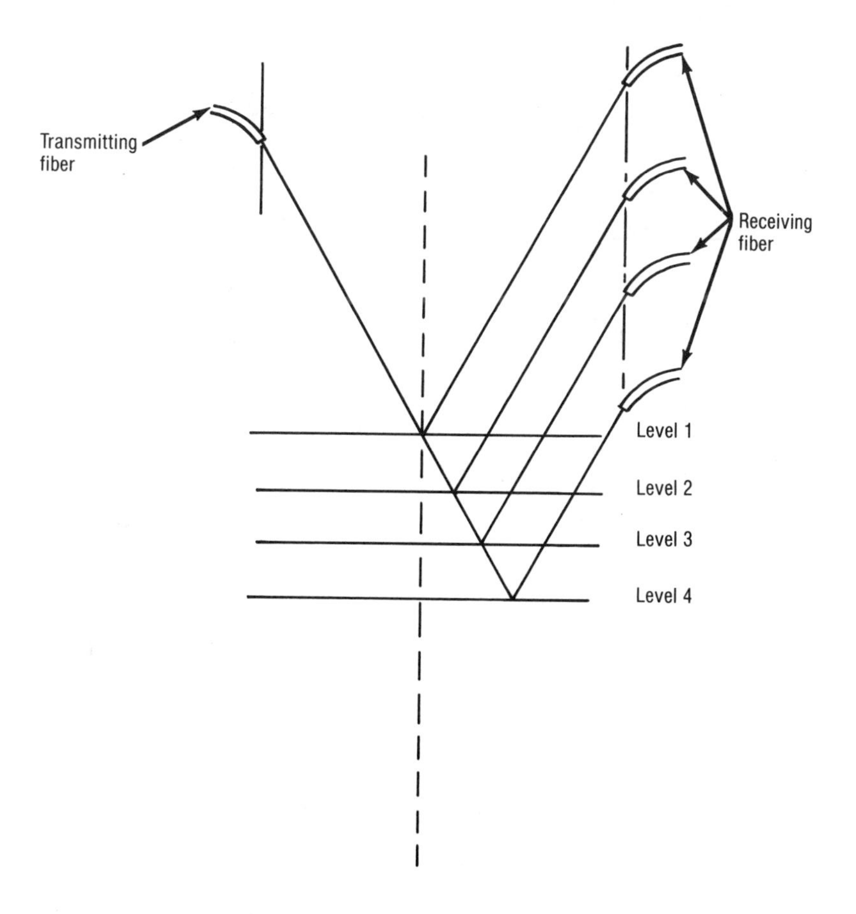

Figure 10-9
Surface Reflectance Liquid Level Measurements

Table 10-1
Output for Various Liquids Using
Fiber Optic/Prism Refractive Index Change Sensor

Medium	Relative Output
Air	1.00
Water	0.11
Isopropyl alcohol	0.07
Gasoline	0.03
Milk	0.20

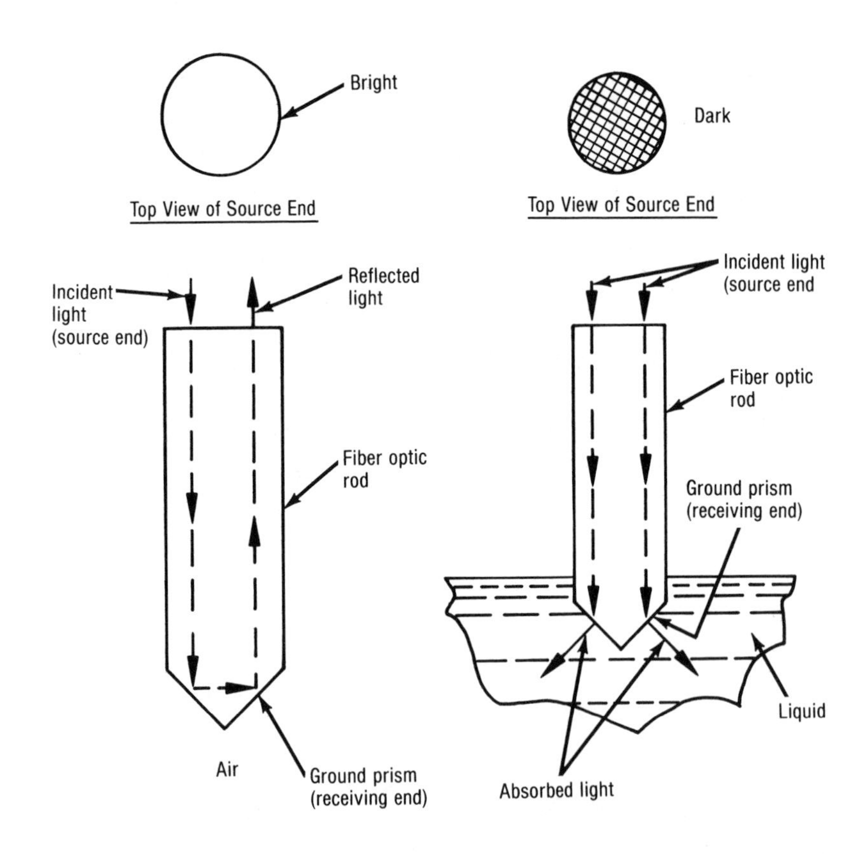

Figure 10-10
Refractive Index Change Liquid Level Sensor

(reprinted by permission from SENSORS, © 1986 Helmers Publishing, Inc.)

An example of such a sensor is for leak detection in a tank farm. The sensor threshold can be set so that water does not trigger the sensor, but gasoline gives an alarm condition.

APPLICATIONS

In describing the various liquid level sensing approaches, several applications have been mentioned. Several other application examples will be discussed to show the broad potential use of fiber optic liquid level sensors.

Volumetric measurement can be achieved if two refractive index change sensors are at different levels in a tank. From the upper level, liquid is drained until the lower level is reached. The difference between the two points is of known volume and can be used for batch processing of liquids.

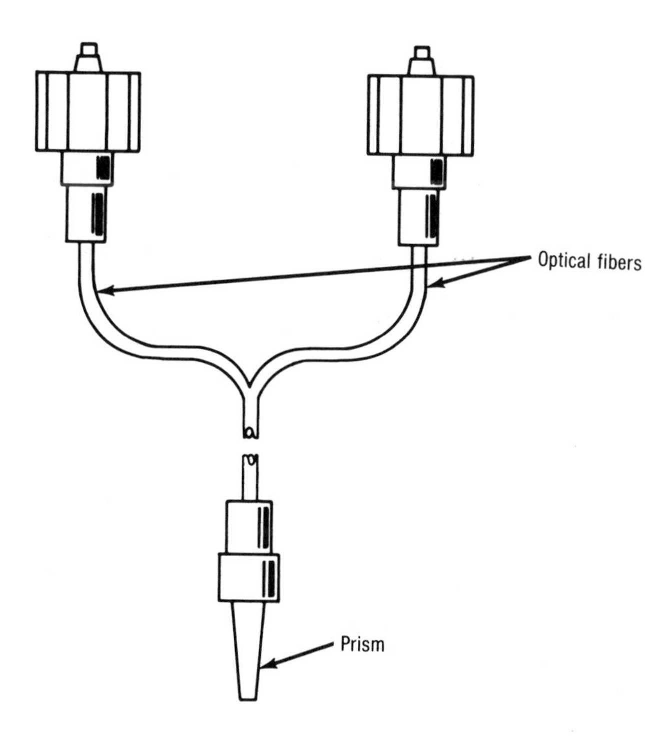

Figure 10-11
Optical Fibers Used to Remote a Prism Type Refractive Index Change Liquid Level Sensor

(reprinted by permission from SENSORS, © 1984 Helmers Publishing, Inc.)

Since the refractive index change sensor requires small amounts of liquid to switch, leaks at lubrication seals can easily be detected. The inverse is also true if the sensor is immersed in a liquid; then, small bubbles can be detected. This technique has use in detecting air leaks in a pipeline or, if attached to a counting system, void fraction in a liquid can be determined.

Most of the fiber optic sensor concepts have applied to point sensors with liquids that are relatively clean and transparent in the UV. Large, dynamic range level magnitude sensors have been limited to pressure-type devices. The surface reflectance technique has found use for high temperature and corrosive liquids in which contact is not possible.

REFERENCES

1. Cho, C. H., 1982, *Measurement and Control of Liquid Level*, ISA, Research Triangle Park, N.C.

2. Gillum, D. R., 1984, *Industrial Level Measurement*, ISA, Research Triangle Park, N.C.

3. Chappel, R. E., and Ballash, R. T., 1983, "Fiber Optics Used for Reliable Drum Water Level Indication," *Engineering Conference — TAPPI Proceedings*, pp 313-317.

4. John, Jr., R. S., 1982, "Fiber Optic Liquid Level and Flow Sensor System," U.S. Patent 4,320,394.

5. King, C., and Merchang, J., 1982, "Using Electro-Optics for Non-Contact Level Sensing," *InTech*, pp 39-40.

6. Rakucewicz, J., 1986, "Fiber Optic Methods of Level Sensing Sensors," *Sensors*, pp 5-12.

7. Coulombe, R. F., 1984, "Fiber Optic Sensor Catching Up with the 1980's," *Sensors*, pp 5-10.

Magnetic and Electric Field Sensors

11

INTRODUCTION

The monitoring of current (magnetic field) and voltage (electric field) are critical for power utilities and in other applications where high electrical power is used. At the time of this writing, sensor technology used in high voltage substations is characterized as reliable but quite expensive. Fiber optic sensor technology is especially attractive since it is immune to electromagnetic interference (a major problem in existing systems) and potentially represents a lower cost alternative.[1,2,3]

Several sensing approaches have been conceived using fiber optics for both magnetic and electric field sensing. The various optical and opto-mechanical effects such as Faraday rotation, Kerr effect, Pockels effect and magnetostriction will be defined later in this chapter. For magnetic field detection, several intensity modulation schemes have been reported using magneto-optical materials to achieve Faraday rotation. A microbending technique has also been demonstrated. Considerable work has been done on magnetostrictive materials used in a phase modulation configuration. For electric field sensing, intensity modulation schemes have been described using electro-optical materials to utilize the Pockels and Kerr effects. Several reflective concepts have been reported using both target and fiber characteristics. Phase modulation techniques for electric field monitoring have centered around piezoelectric fiber coatings.

MAGNETIC FIELD

Intensity Modulation

A transmissive fiber optic sensor can be used to measure the magnetic field and to determine the current if a magneto-optical material is placed in the light path.[5,6,14] The material can be incorporated intrinsically in the fiber or a bulk material can be inserted between two fibers in a transmissive mode as shown in Figure 11-1.

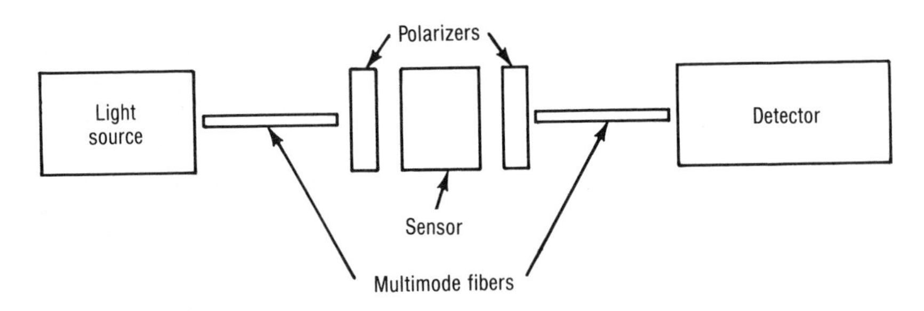

(a) Bulk Sensor Material

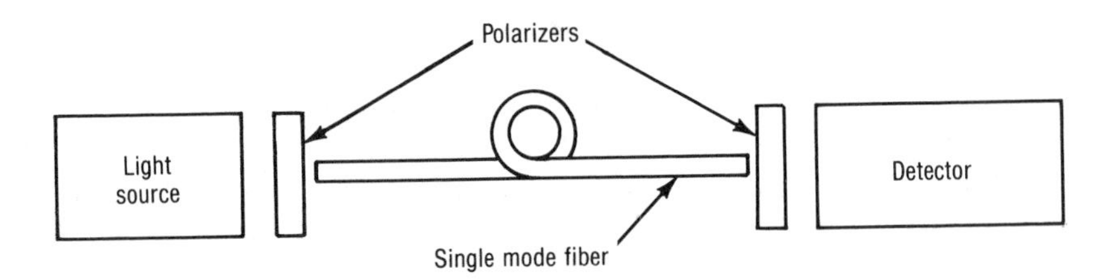

(b) Intrinsic Sensor

Figure 11-1
Fiber Optic Transmissive Magnetic Field Sensor Using Magneto-optic Sensing Material in the Light Path

(© 1985 IEEE; reprinted with permission)

Magneto-optic materials use the Faraday effect, which produces a change in refractive index as described in the following equation:[2]

$$n_R - n_L = \lambda VH/\pi \tag{11-1}$$

where n_R is the index of refraction for right circulary polarized light; n_L is the index of refraction for left circularly polarized light; λ is the wavelength of the light beam; H is the magnetic field intensity; and V is the Verdet constant. In essence, in the presence of a magnetic field, the plane of polarization is rotated, which directly modulates the intensity of the transmitted beam in a manner proportional to the field. The equation for the rotation of the polarized beam, θ, is:[4]

$$\theta = VHl \tag{11-2}$$

where l is the length of magneto-optical material in the field. The transmitted intensity is proportional to $\cos^2\theta$ or $\cos^2(VHl)$.

The distinct advantage of incorporating the Faraday effect in the fiber is that the sensor has the potential for a much longer light path than that of bulk optical material. As defined in Equation 11-2, the increased light path directly increases the sensitivity of the Faraday rotator. A problem with the intrinsic fiber sensor is the interaction of the fiber itself with the polarization associated with the magnetic field. One interaction is residual birefringence in the fiber associated with strains induced by drawing and coiling. To maximize sensitivity, low birefringence fiber must be used.[28] Also, the fiber tends to depolarize the light over a large path length. Multimode fibers with multiple-mode interactions tend to be much greater depolarizers than single-mode fibers. Single-mode fibers with low residual birefringence are used in sensors of this type.

The Faraday effect in diamagnetic materials, such as pure silica, is relatively temperature-insensitive; however, the Verdet constant is low. The doping of silica with rare earth ions such as terbium can significantly increase the Verdet constant, but the addition of these paramagnetic materials will also introduce a temperature dependence.

Using a fiber optic transmissive sensor with a Faraday material as the sensing element, currents in the range of 100 to 15,000 amperes have been measured.[5] The sensor accuracy is reported to be 1%. The linearity of the sensor response is limited to a relatively small portion of the dynamic range of the sensor.

A technique is described for measuring relatively low currents using a metal coated fiber and monitoring microbending effects.[7] The fiber is coated with aluminum. When current is passed through the coating and the fiber (coiled on a serrated mandrel) is in magnetic field, the magnetic force on the fiber produces bending. The bending modulates the speckle pattern, which was discussed in Chapter 9. The sensor response versus current is plotted in Figure 11-2. Linearity was observed between 5 and 2000 mA. The threshold for detection by this technique appears to be 5 mA.

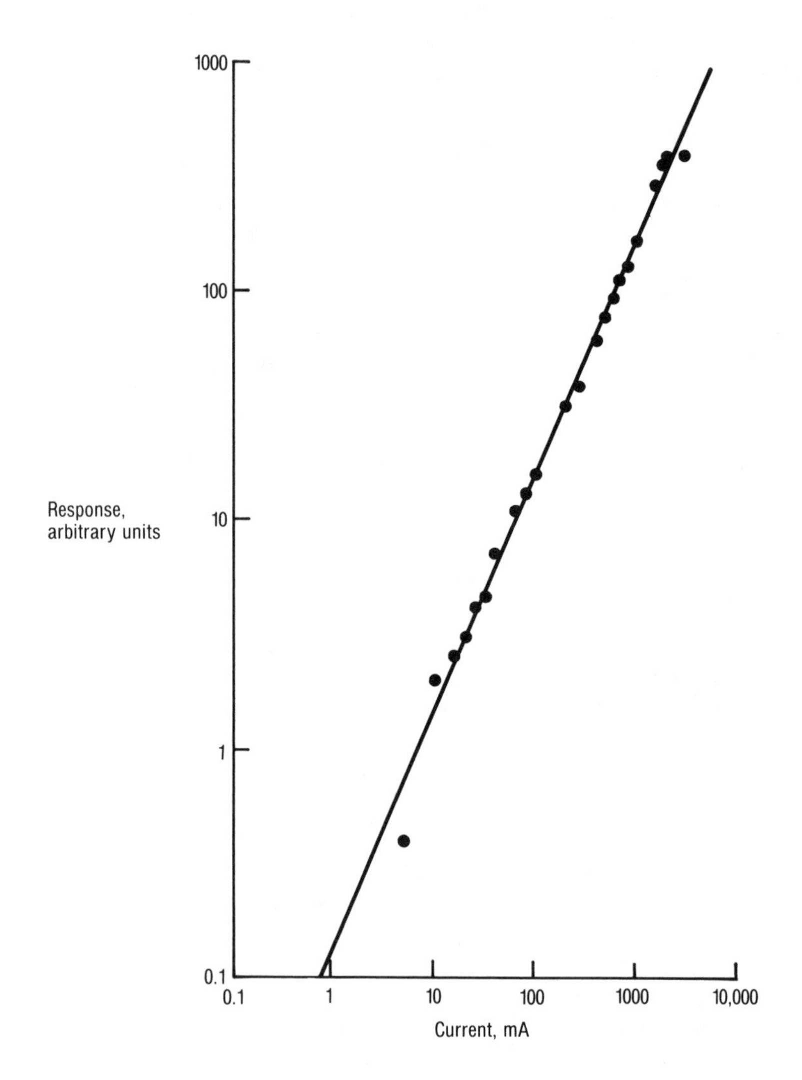

Figure 11-2
Response Curve for a Microbending Magnetic Field Sensor

(© 1980 IEEE; reprinted with permission)

Phase Modulation

Phase modulation techniques for measuring magnetic fields have been used primarily in conjunction with magnetostrictive materials.[8-13] Magnetostrictive materials expand or contract, depending upon the material, in the presence of a magnetic field. Figure 11-3 depicts the geometry of the coating on a single-mode silica fiber. Ideally, the magnetostrictive coating is in a fiber form, but bulk materials can be used such as stripes or cylinders with the fiber bonded to the material as shown in Figure 11-4. Several available alloys exhibit magnetostrictive properties, such as nickel alloys, cobalt-iron alloys, and metallic glasses. Figure 11-5 illustrates positive and negative magnetostriction with various alloys.

The sensor uses a Mach-Zehnder configuration with the magnetostrictive material perturbing the sensing fiber. The basic configuration is shown in Figure 11-6. There is an optimum bias condition to achieve maximum sensitivity. The dc magnetic field bias can increase sensitivity by a factor of four.

Using a fiber bonded to a metallic glass strip, Koo and Sigel[9] demonstrated the linear sensor response in an ac magnetic field at 1 kHz, as shown in Figure 11-7. The dc magnetic field bias and strip length are indicated. The response was linear over a field strength variation of two orders of magnitude. Others[12] have reported a linear response over six orders of magnitude.

The frequency response of a phase-modulated magnetostrictive fiber optic sensor is very dependent upon the material as well as the geometry. However, above approximately 1 kHz, a general decrease in sensitivity is observed.

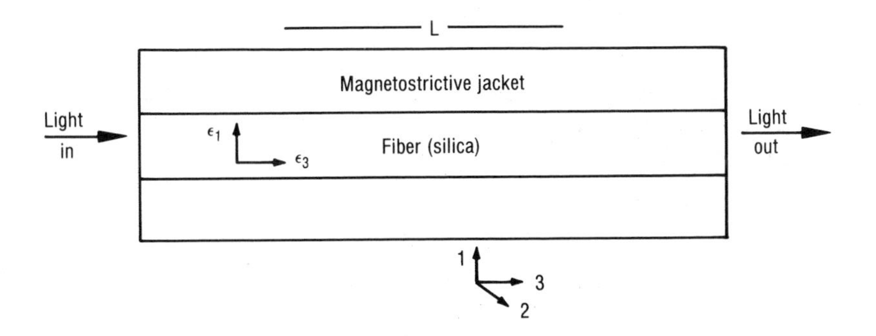

Figure 11-3
The Basic Geometry of a Silica Fiber Embedded in a Magnetostrictive Jacket

(reprinted by permission from the Optical Society of America)

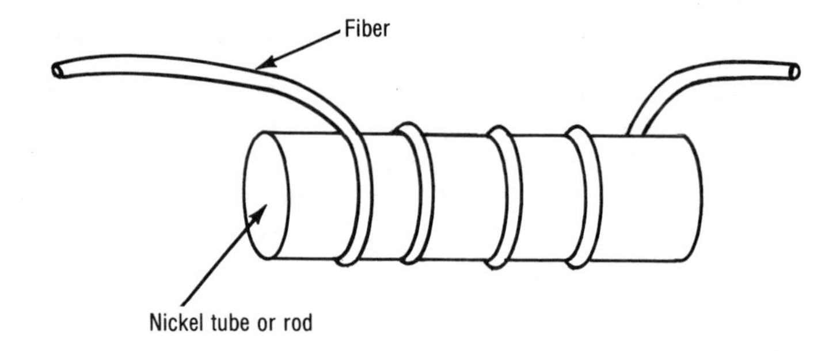

(a) Mandrel Sensor

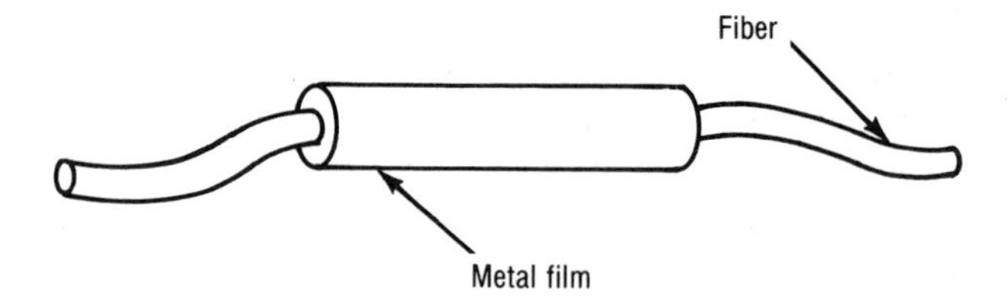

(b) Coated Sensor

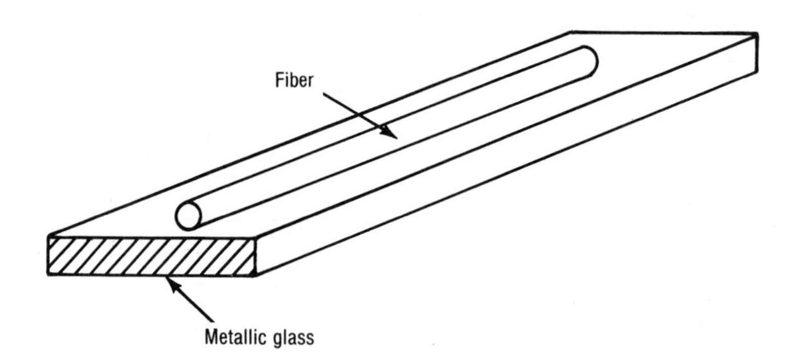

(c) Stripline Sensor

Figure 11-4
Basic Configurations of Magnetic Fiber Sensors

(© 1982 IEEE; reprinted with permission)

The work of Koo and Sigel[9] resulted in very high sensitivity magnetic field sensors. The minimum detectable field is approximately 5×10^{-12} Oe/m.[13]

ELECTRIC FIELD

Intensity Modulation

In a manner similar to an electric field, a transmissive fiber optic sensor can be used to measure an electric field if an electro-optic material is placed in the light path.[2,3,14,15] Unlike magneto-optical materials, there are no components that have been effectively added to a glass fiber to achieve intrinsic sensing. The sensors reported in the literature use electro-optic crystals.

Electro-optic materials generally are classified as exhibiting the Pockels effect or the Kerr effect. For the Pockels effect in the presence of an electric field, the index of refraction change is directly proportional to the field.

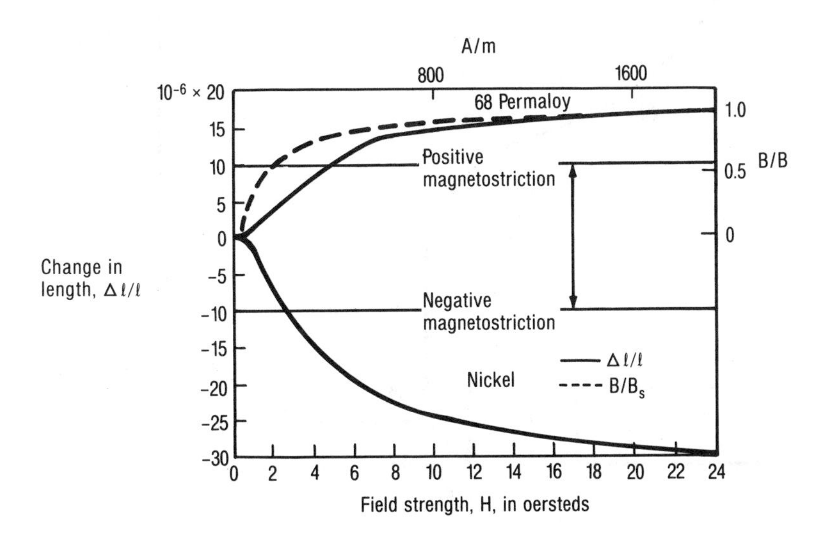

Figure 11-5
The Magnetostriction of Nickel and of 68 Permalloy
(reprinted by permission from the Optical Society of America)

Materials that exhibit the Kerr effect have an index change that is proportional in the square of the field. The Kerr effect is more prevalent in liquids. A variety of crystals exhibit the Pockels effect,[16] and for this reason most experimental work on the concept relates to the Pockels sensor.

Figure 11-8 shows a transmissive Pockels sensor. The incoming light is plane-polarized. In the presence of an electric field, the orthogonally polarized beams propagate with different velocities due to the refractive index difference associated with the Pockels effect. A phase difference is generated,

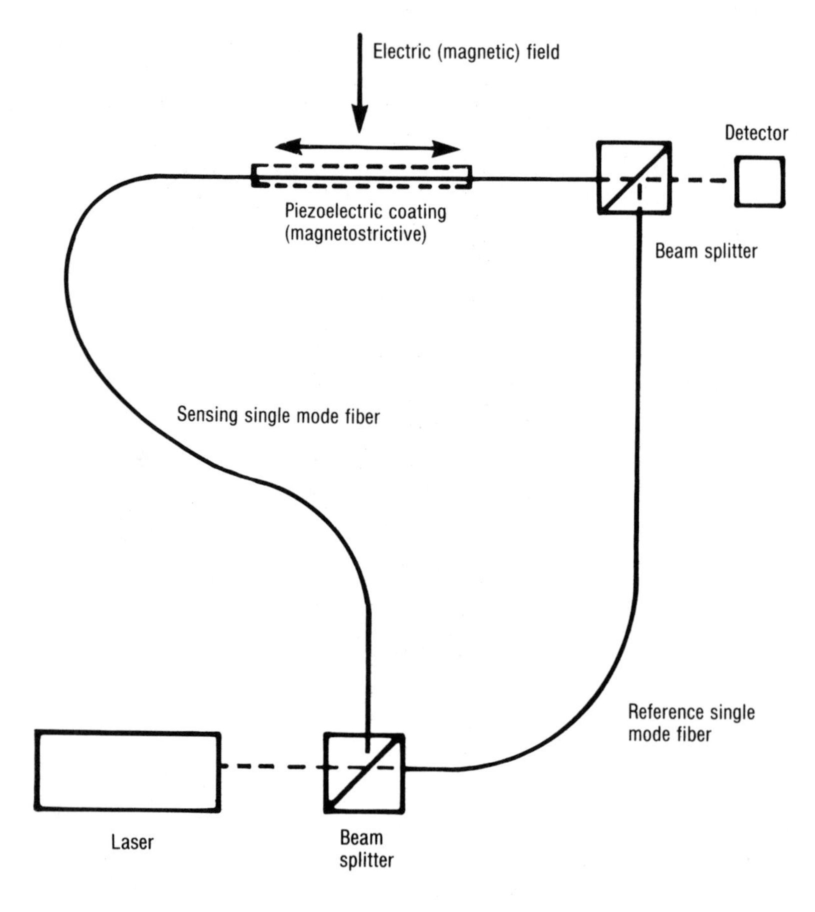

Figure 11-6
Mach-Zehnder Sensor Configuration for Electric or Magnetic Field Sensing

which, with the aid of a biasing quarter wave plate, rotates the plane of polarization as a function of voltage. The transmitted intensity, I, is given by:[14]

$$I = 1/2\, I_o\, (1 + \sin V/V_0) \qquad (11\text{-}3)$$

where I_o is the input light intensity, V_0 is the half wave voltage of the Pockels material (a property of the material), and V is the applied voltage.

Figure 11-9 shows a plot of sensor output versus applied voltage for a transmissive Pockels cell sensor using $Bi_{12}GeO_{20}$ as the electro-optic material. The results were linear over nearly three orders of magnitude.[14]

Several techniques for measuring electric field or voltage using a reflective technique are reported in the literature.[17,18,19] Figure 11-10 shows a schematic representation of a voltage sensor. The sensor uses a reflective probe that is quite sensitive to the motion of the reflective target, which is a mirror attached to a piezoelectric element. In the presence of an electric field, the piezoelectric element moves, causing the mirror to move and alter the reflected light signal in a manner proportional to the applied voltage.

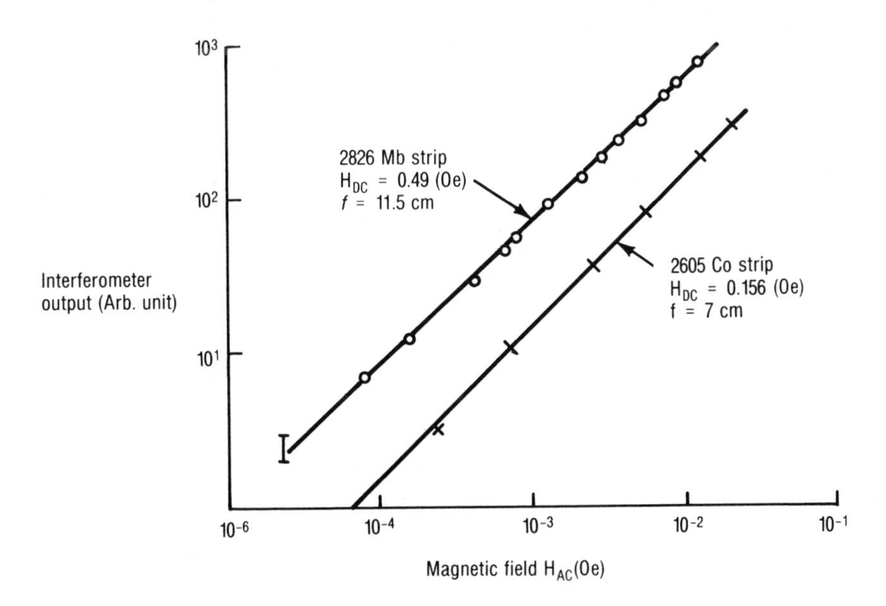

Figure 11-7
Linearity of Interferometer Responses versus Magnetic Fields
(reprinted by permission from the Optical Society of America)

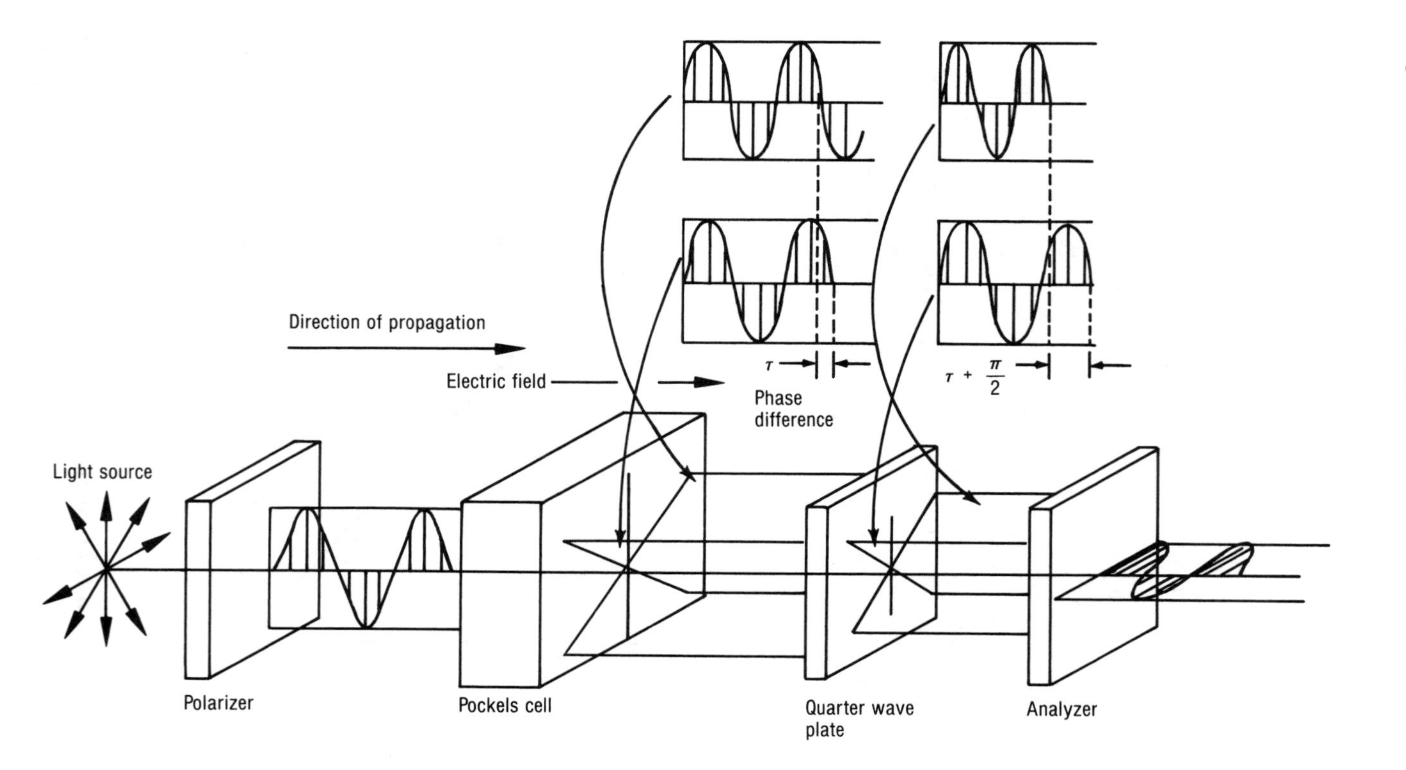

Figure 11-8
Fiber Optic Transmissive Electric Field Sensor Using a Pockels Cell

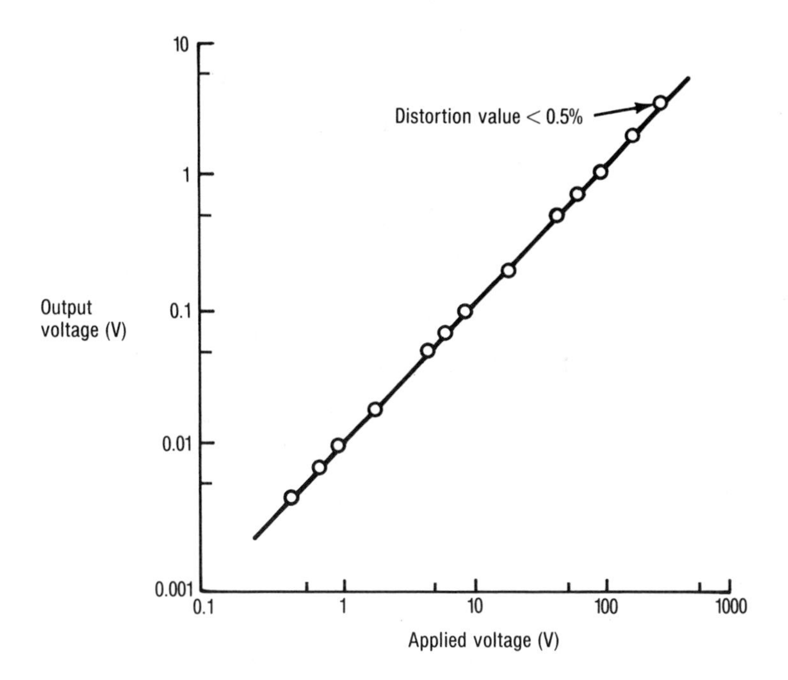

Figure 11-9
Linearity Between Applied Voltage and Output Voltage for Pockels Cell Sensor ($Bi_{12}Ge_{20}$)

Figure 11-11 shows a fiber optic electroscope.[(18)] The fibers have a thin conductive coating. In response to an electric field, the fibers bend away from each other due to an unbalanced lateral force on each fiber that is proportional to the square of the field. The fibers are always deflected outward as the field increases. The reflected light, therefore, decreases with an increase in field strength.

Another approach is the incorporation of a pressure-sensitive diaphragm in the wall of an electrode.[(19)] In the presence of a field, forces are exerted on the diaphragm. The diaphragm deflection, which is monitored fiber optically, is proportional to the square of the field (as with the electroscope described above).

Phase Modulation

Phase modulation techniques for measuring electric fields have been used with electromechanical materials, primarily piezoelectric materials.[(20,21,22,23)]

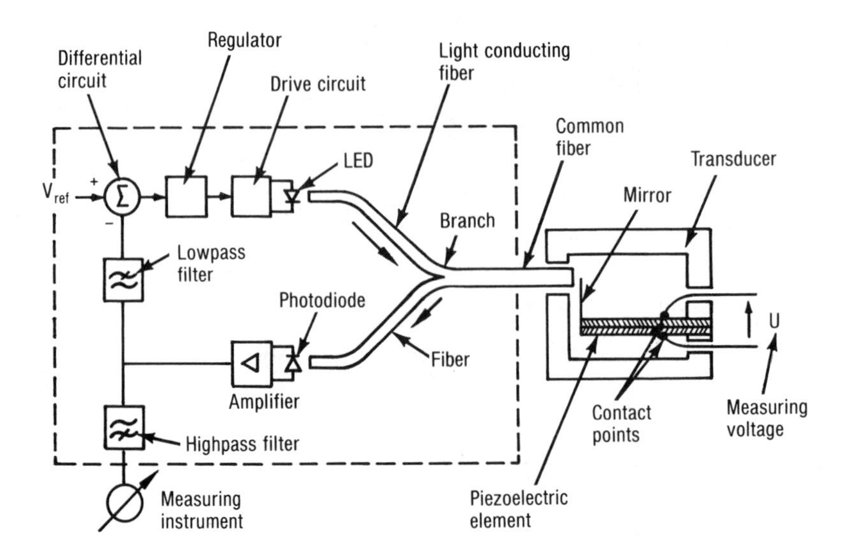

Figure 11-10
Fiber Optic Reflective Sensor for Measuring Voltage Using a Piezoelectric Element

(patented)

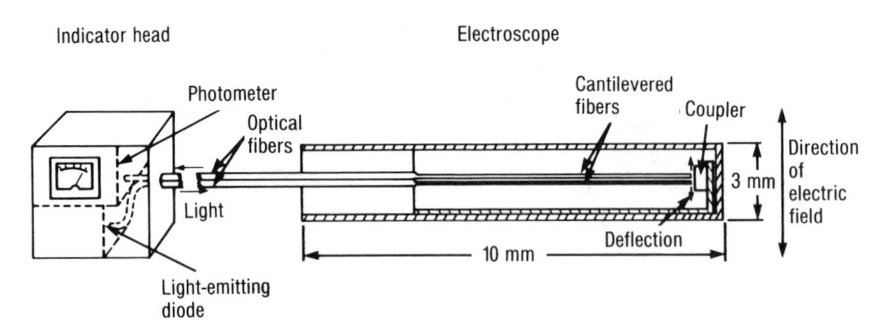

Figure 11-11
Fiber Optic Electroscope Using a Reflective Sensor

In a manner similar to magnetostrictive materials in a magnetic field, piezoelectric materials can expand and contract in the presence of an electric field. Piezoelectric materials interact with the sensing fiber in a Mach-Zehnder interferometer by being bonded to the fiber. The configuration can be a coated fiber, a fiber bonded or embedded in a sheet of the material, or a fiber wrapped around a mandrel. Most of the reported experimentation has been done with polyvinylidene fluorine (PVF_2) as the piezoelectric material.

The interferometric configuration is shown in Figure 11-12.[22] The response versus applied field is shown in Figure 11-13. The fiber was coated with a PVF_2 coating, which was poled to optimize the molecular orientation in the polymer and maximize the piezoelectric effect. The curve indicates that the response is linear. For this particular response curve, the coating thickness was about 25 microns. DeSouza and Mermelstein[23] found that the sensitivity increases with coating thickness up to about 200 microns. Coatings thicker than 200 microns have little added advantage. The projected minimum detectable field is 4 μV/m for a sensor with a 1 km sensing fiber length.

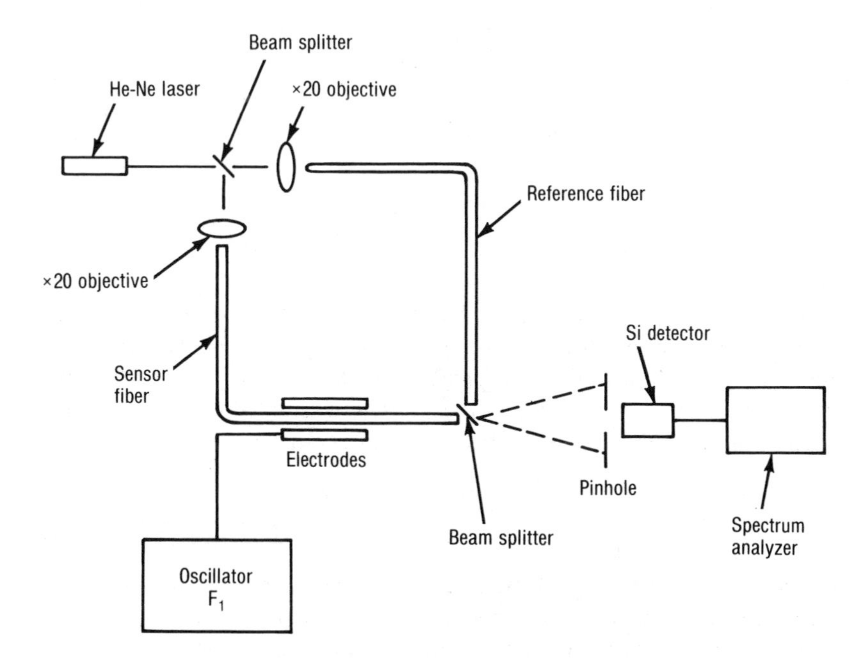

Figure 11-12
Schematic Diagram of Measurement System

(reprinted by permission from the Optical Society of America)

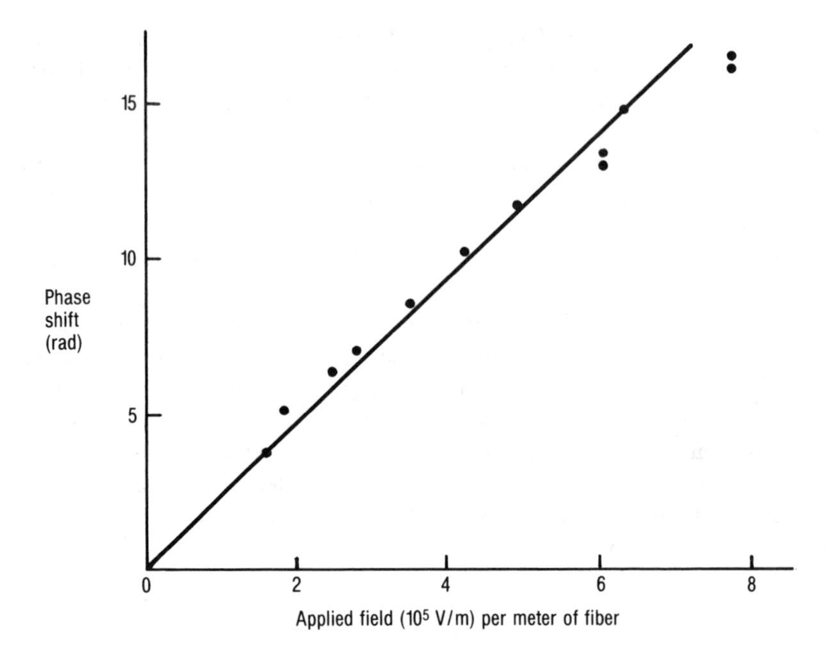

Figure 11-13
Observed Phase Shift as a Function of Applied Electric Field

Koo and Sigel[21] using a fiber bonded to a PVF_2 strip found that the interferometric output was essentially constant up to frequencies of 1 kHz. Beyond this level, deviations occurred in the form of both maximum and minimum points. The nonuniform response is attributed to resonance of a mechanical nature associated with the sensor geometry.

APPLICATIONS

The primary applications of magnetic and electric field sensors are in the utility area as discussed at the beginning of this chapter. Examples of other areas of potential application are high energy plasma diagnostics, electromagnetic signal detection, and magnetic heading determination via a magneto-optic compass.

REFERENCES

1. Malewski, R., and Erickson, D., 1985, "Fiber Optic Applications in Electrical Power Systems, Measurement Applications (Part 1)," *IEEE Tutorial Course*, IEEE, Piscataway, N.J.

2. Habner, R. E, 1985, "Fiber Optic Applications in Electrical Power Systems, Measurement Application in Electrical Power Systems, Measurement Application (Part 2)," *IEEE Tutorial Course,* IEEE, Piscataway, N.J.

3. Habner, R. D., 1985, "Optical Technology Applications in Gas and Electric Utilities," Presented at the Newport Conference on Fiber Optic Markets.

4. Borrelli, N. F., Kapnon, F. P., and Keck, D. B., 1973, "Optical Waveguide Light Modulator," U.S. Patent 3,756,690.

5. Anonymous, 1980, "Lasers and Fibers Monitor Current and Voltage at High Power Stations," *Laser Focus*, pp 48–56.

6. Ullrich, R., 1981, "Fiber Optical Arrangement for Measuring the Intensity of an Electric Current," U.S. Patent 4,255,018.

7. Tangonan, G. L., Persechini, D. I., Morrison, R. J., and Wysocki, J. A., 1980, "Current Sensing with Metal-Coated Multimode Optical Fibers," *Electronic Letters*, Vol. 16, No. 25, pp 958–959.

8. Yariv, A., and Winsor, H. V., 1980, "Proposal for Detection of Magnetic Fields through Magnetostrictive Perturbation of Optical Fiber," *Optics Letters*, Vol. 5, No. 3, pp 87–89.

9. Koo, K. P., and Sigel, G. H., 1982, "Characterizations of Fiber-Optics Magnetic Fields Sensors Employing Metallic Glasses," *Optics Letters*, Vol. 7, pp 334–336.

10. Dandridge, A., Tvetan, A. B., Sigel, G. H., West, E. J., and Giallorenzi, T. G., 1980, "Optical Magnetic Field Sensors," *Electronic Letters*, Vol. 16, pp 408–409.

11. Jarzynski, J., Cole, J. H., Bucaro, J. A., and Davis, C. M., 1980, "Magnetic Field Sensitivity of an Optical Fiber with Magnetostrictive Jacket," *Applied Optics*, Vol. 19, No. 22, pp 3746–3748.

12. Giallorenzi, T. G., et al., 1982, "Optical Fiber Sensor Technology," *IEEE Journal of Quantum ELectronics*, Vol. QE-18, No. 4, pp 626–665.

13. Davis, C. M., et al., 1982, *Fiber Optic Sensor Technology Handbook*, Dynamic Systems, Reston, Virginia.

14. Kuroda, Y., et al., 1985, "Field Test of Fiber Optic Voltage and Current Sensor Applied to Gas Insulated Substation," *Proceedings SPIE, Fiber Optic Sensors*, Vol. 586, pp 30–37.

15. Fernandes, R. A., 1984, "Fiber Optic Electric Utility Applications Present and Future," *Fiber Optic Applications in Electrical Utilities*, IEEE, Piscataway, N.J.

16. Driscoll, W. G., and Vaughan, W. (Eds.), 1978, *Optics Handbook*, McGraw-Hill, New York, pp 10–141.

17. Adolfsson, M., and Brogradeh, T., 1986, "Fiber Optic Sensor for Measuring Current or Voltage," U.S. Patent 4,547,729.

18. Johnson, A. R., Fall 1985, "Fiber Optic Electric Field Meter," *NASA Tech. Briefs*, pp 39–40.

19. Byberg, B. R., Herstad, K., Larsen K. B., and Hansen, T. E., 1979, "Measuring Electric Field by Using Pressure Sensitive Elements," *IEEE Transactions on Electrical Insulation*, Vol. EI-14, No. 5, pp 250–254.

20. Mermelstein, M. D., "Optical Fiber Copolymer-Film Electric-Field Sensor," *Applied Optics*, Vol. 22, No. 7, pp 1006–1009.

21. Koo, K. P., and Sigel, Jr., G. H., 1982, "An Electric Field Sensor Utilizing a Piezoelectric Polyvinylidene Fluoride (PVF_2) Film in a Single-Mode Fiber Interferometer," *IEEE J. Quantum Electronics*, Vol. QE-18, No. 4, pp 670–675.

22. Donalds, L. J., French, W. G., Mitchell, W. C., Swinehart, R. M., and Wei, T., 1982, "Electric Field Sensitive Optical Fiber Using Piezoelectric Polymer Coating," *Electronics Letters*, Vol. 18, No. 8, pp 327–328.

23. DeSouza, P. D., and Mermelstein, M. D., 1982, "Electric Field Detection with a Piezoelectric Polymer-Jacketed Single-Mode Optical Fiber," *Applied Optics*, Vol. 21, No. 23, pp 4214–4218.

24. Anonymous, 1984, "Fiber Sensors Impact on Power Utilities," *Lightwave*, p. 12.

25. Kyuma, K. et al., 1983, "Fiber Optic Current and Voltage Sensors Using a $Bi_{12}GeO_{20}$ Single Crystal," *Journal of Lightwave Technology*, Vol. LT-1, No. 1, pp 93–97.

26. Day, G. W., Veeger, L. R., and Gernosek, R. W., March 1986, "Progress in the Design of Optical Fiber Sensors for the Measurement of Pulsed Electric Currents," *Proc. of the Workshop on Measurement of Electrical Quantities in Pulse Power Systems*, Gaithersburg, MD.

Chemical Analysis

12

INTRODUCTION

Fiber optic techniques for chemical analysis have several distinct advantages.[1] Analysis can often be done *in situ* in real time. The sensing techniques generally do not disturb the process. The sample size can be extremely small, and the sensing locations can be in remote areas that are normally difficult to access. Potential disadvantages include sensitivity to ambient light, relatively slow response time due to the required reaction with various reagents, and shortened lifetime if high incident radiation is used to enhance sensitivity.

Four approaches can be used for qualitative and quantitative chemical analysis. These techniques include fluorescence, scattering, absorption, and refractive index change.

In general, chemical analysis techniques employ either a transmissive or a reflective fiber optic configuration. Figure 12-1 illustrates the two concepts. The reflective concept uses a bifurcated probe, as shown in Figure 12-1 (a). Light travels down the transmitting leg, reflects off the target material, and is accepted in the receiving leg, which is attached to a photodetector. The amount of light transmitted or reflected is a function of the nature and amount of chemical species present. In the transmissive system shown in Figure 12-1 (b), light travels down a transmitting fiber optic (a single fiber or a bundle of many fibers), passes through a gap that contains the material to be analyzed, and is captured in a receiving fiber optic, which, in turn, transmits the light to a photodetector.

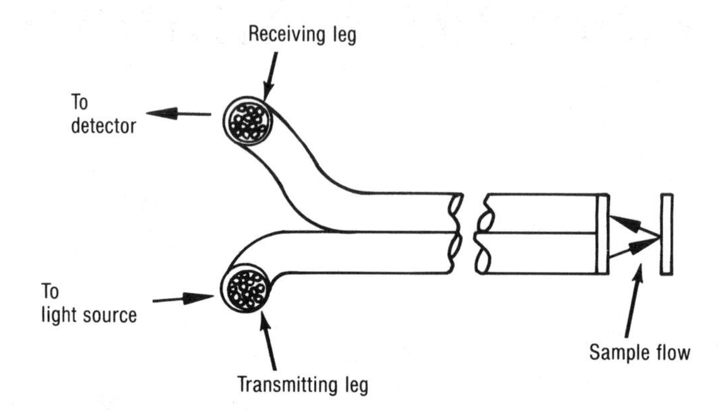

(a) Reflective System

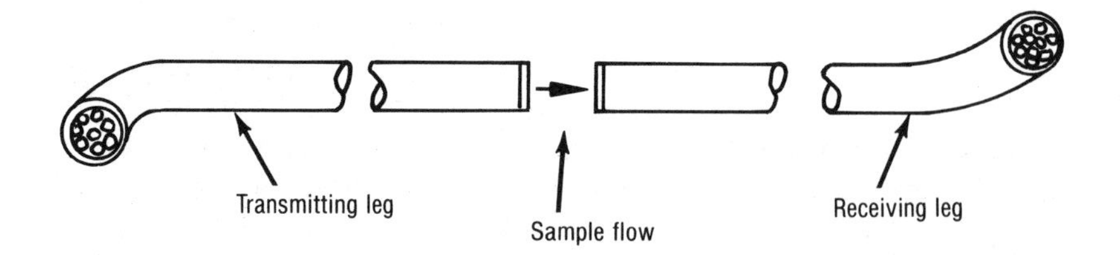

(b) Transmissive System

Figure 12-1
Fiber Optic Sensing System

FLUORESCENCE

The use of fiber optic fluorescence techniques for chemical analysis is generally referred to as remote fiber fluorimetry (RFF).[2,3,4,5] A typical RFF system couples a high intensity light source into a single, large core quartz fiber. Light travels along the length of the fiber as a result of total internal reflection, with little optical loss. Upon exiting the fiber, the rays of light impinge upon the sample, which, in turn, gives off a characteristic fluorescent emission. The emission is detected by the same fiber and travels back to a photodetector, which, with the aid of a computer, can provide both qualitative and quantitative information (see Figure 12-2). The fiber immersed in the sample is called an optrode. The fluorescence can be enhanced with the aid of a sapphire microlens, as shown in Figure 12-3.

Using the RFF technique, measurements have been demonstrated with uranyl, chloride, iodide, iron, plutonium, and sulfate ions.[4] Ground water contamination, such as toluene and xylene associated with gasoline spills and leakage,[6] has also been detected by the technique.

Hirschfeld[2] has pointed out that a material does not have to be fluorescent to work with the technique. A target of known fluorescence can be used. The sample in question can react with the target and enhance or diminish its fluorescence. Aluminum and other metals have been analyzed by using a reagent immobilized in the form of a powder and attached to a bifurcated fiber optic probe.[7] The metal reacts with the reagent, giving a fluorescent signal. The response time was 1 to 2 minutes with a detection limit of 0.027 ppm.

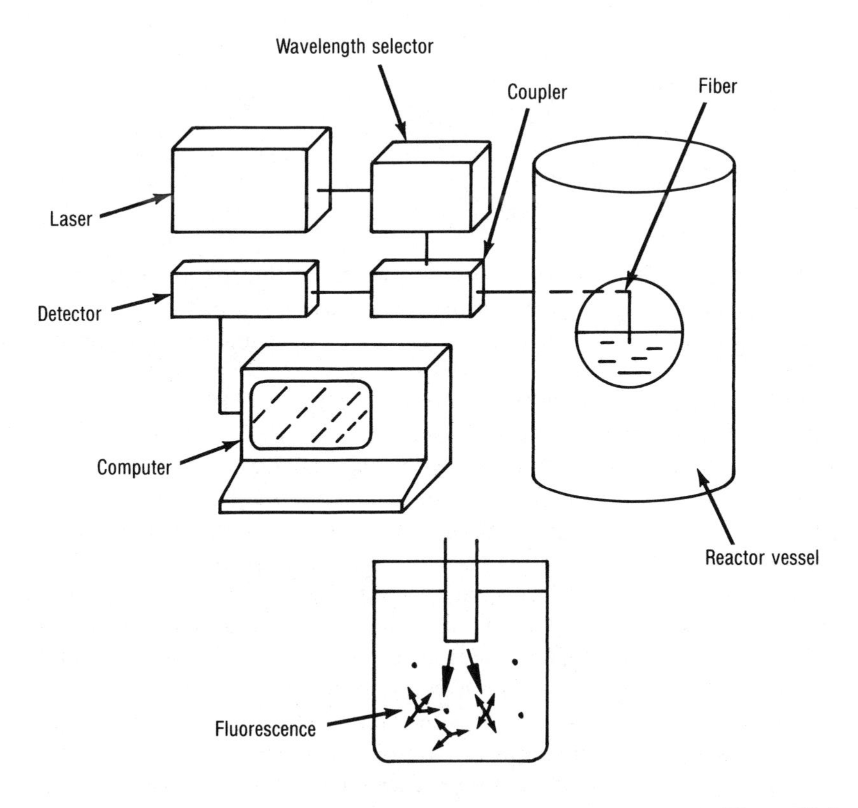

Figure 12-2
Remote Fiber Fluorimetry

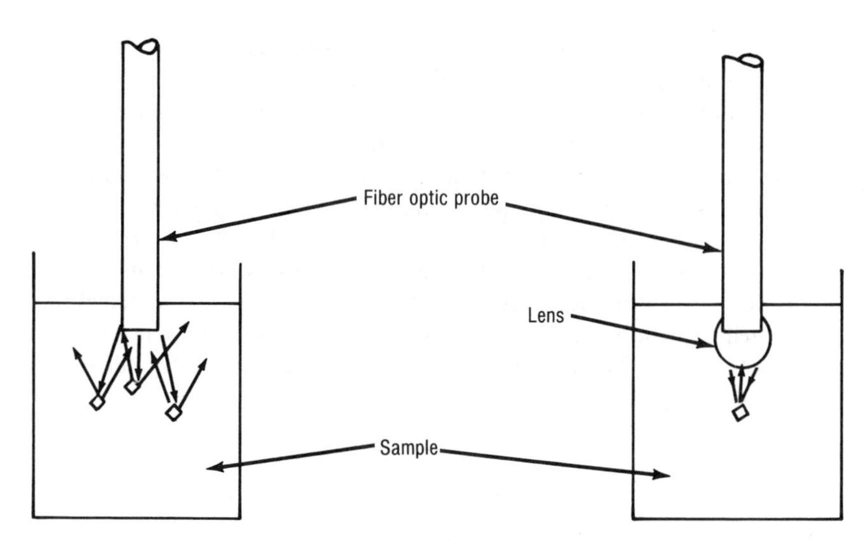

(a) Unfocused System *(b) Focused System*

Figure 12-3
Unfocused and Focused Remote Fiber Fluorimetry Probes

Oxygen partial pressure has been measured by a fluorescent technique.(8) The technique uses a fluorescent dye that is quenched in the presence of oxygen. The device is shown in Figure 12-4. A high intensity blue light is transmitted to the dye through one leg of a bifurcated fiber optic probe. The blue light, upon impinging the dye, gives off a characteristic green fluorescence. The level of fluorescence diminishes with the increasing levels of oxygen that pass through the gas-permeable membrane, which, in turn, reacts with the dye. The partial pressure of oxygen is a function of the ratio of blue light intensity to green light intensity. Over a range of 0 – 150 torr, the sensor was accurate to better than 1% with a response time of about 2 minutes. Seitz(9) has used a similar technique for measuring the concentration of glucose. Halide contaminations such as iodide, bromide, and chloride were also detected using fluorescence quenching.(10) The detection limit was about 2 ppm for iodide, 6 ppm for bromide, and 200 ppm for chloride.

The fluorescence technique has been expanded to measure pH. Saari and Seitz,(11) using a fiber optic bifurcated probe in conjunction with an immobilized reagent, found that the fluorescence intensity changed with pH and/or the concentration of metal ion. The apparatus is shown in Figure 12-5. The device had good sensitivity over the pH range of 3 to 8.

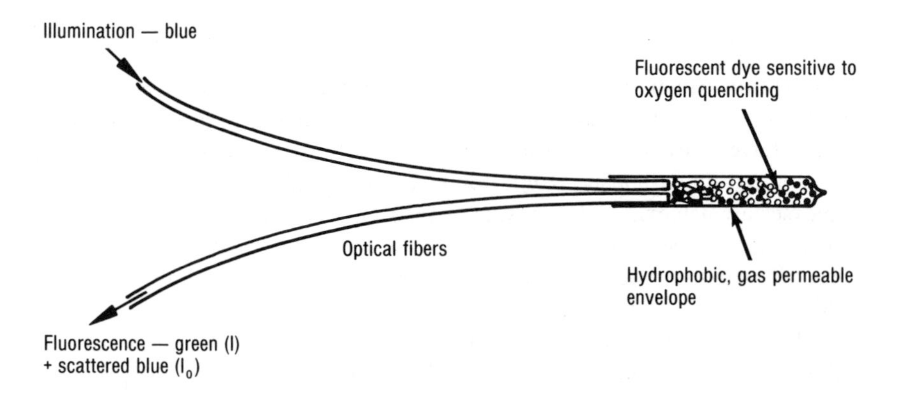

Figure 12-4
Construction of Fiber Optic Fluorescent Probe

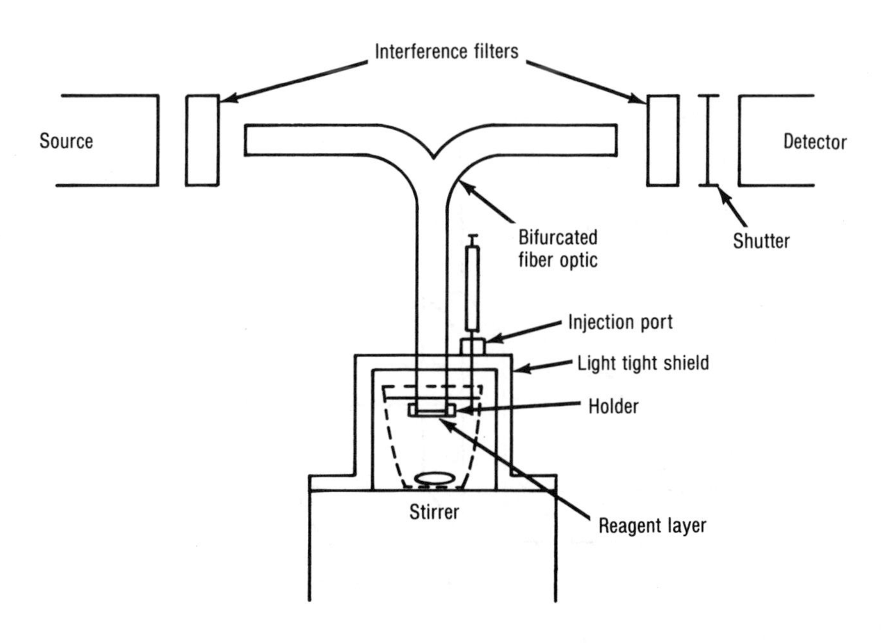

Figure 12-5
Diagram of pH Sensor Based on Fluorescence and Associated Instrumentation

ABSORPTION

Absorption can be used as an analysis approach.[4] Figure 12-6 shows a reflective system in which a gas sample to be measured is in the optical path. Gases have characteristic absorption bands. If the fiber optic sensor is used in conjunction with a spectrophotometer, both qualitative and quantitative analyses are achievable. For a weakly absorbing gas, the light path through the gas may have to be long to enhance the sensitivity required. This problem can usually be solved by use of a faceted target, which allows multiple reflections through the material being analyzed.

There is a major limitation in any fiber optic system. The best quality fiber can transmit only between the UV to mid IR; therefore, the primary absorption peaks of many gases are beyond the range of present fiber optic systems. However, many gases have weaker overtone absorption bands in the near IR that can be used in analysis.[4,12,13]

While the use of such techniques is in elementary stages, the same device, shown in Figure 12-6, is being considered for use in a scattering mode for smoke detection (switch) in both military and industrial applications.

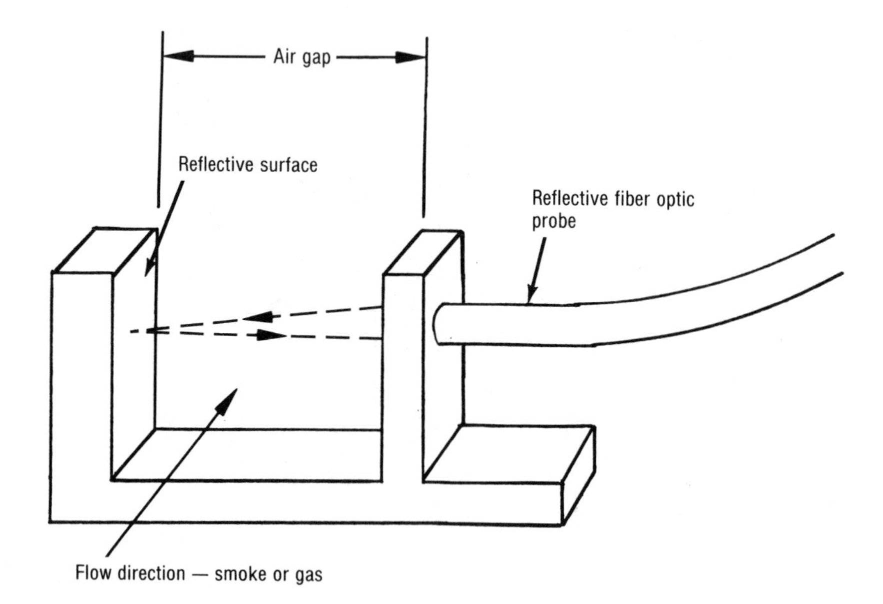

Figure 12-6
Gas/Smoke Sensor

The usefulness of absorption-based fiber optic sensors can be enhanced if reagents that react with the substance to be detected are used.[9,14,15] The reaction can be designed to occur within the operational wavelength range of the sensor, even if the substance has no characteristic absorption bands within the operational wavelength range. Figure 12-7 shows a configuration of such a sensor. The reagent is immobilized, which entraps the reagent and allows the sensor to function in a reversible manner. The sensor design allows the analyte to diffuse through the membrane surrounding the reagent. The reagent is typically a dye that changes its absorption as a function of concentration. At a given wavelength, the reflected light is diminished in intensity as a function of analyte concentration. The sensor concept is further enhanced if a second wavelength that does not attenuate the light level is used. The intensity associated with the nonabsorption wavelength is used as a reference to normalize the intensity associated with absorption wavelength. It has been shown that the reagent can be in the light path of a reflective-type sensor or a transmissive sensor. However, it is also possible to have the reaction occur along the wall of the fiber since some of the light penetrates beyond the wall (evanescent wave).[9] The reaction can be enhanced if a short section of fiber has the cladding removed and the reagent is in direct contact with the core, as shown in Figure 12-8. The distinct advantage of this approach is that the sensor has lower inherent losses compared to a reflective or a transmissive sensor, since the light path is unbroken and loss at the sensing point is minimized (but detectable). This approach also has an advantage in a distributive sensor configuration, which will be discussed in Chapter 14.

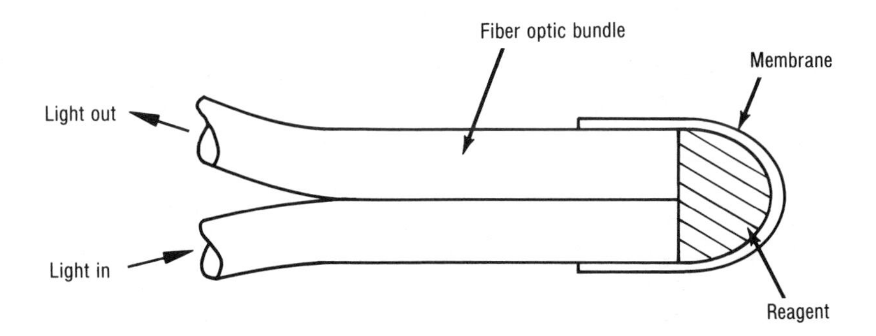

Figure 12-7
Reflective Fiber Optic Probe Using an Immobilized Dye for Absorption Based Sensing

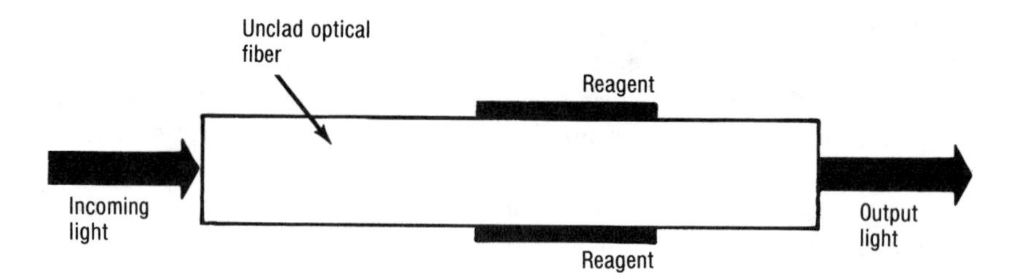

Figure 12-8
Transmissive Fiber Optic Sensor with Reagent Directly in Contact with the Core Along its Walls

A reversible sensor capable of detecting toxic gases was developed using a reagent on the side wall of a waveguide.(16) The reagent was an immobilized dye that changed color in the presence of the gas. Specifically, ammonia vapors were detected to levels of 60 ppm. The response time to reach 90% of the intensity for a given concentration was approximately 1 minute. When the gas was removed, the sensor relaxation time to the zero concentration level was approximately 20 seconds.

Fiber optic colorimetry in its simplest form uses a basic absorption/transmission concept.(23) Many processes, such as antifreeze manufacture, monitor color at various process stages. Figure 12-9 shows a basic colorimeter. The device uses a bifurcated fiber optic reflective probe and a mirror target. The material to be measured is between the probe tip and the mirror. By using specific wavelength interference filters on the receiving leg of probe, specific colors can be monitored. More detailed color characterization is also possible, providing CIE standard color coordinates, but a more complicated sensor configuration is required.

SCATTERING

Many applications require the measurement of the opacity associated with airborne particulate matter. For an opaque particulate, the loss in transmission is due to scattering. For a particulate that is semi-transparent, the loss in transmission is due to both scattering and absorption. Opacity is defined as follows:(17)

$$\%\ \text{opacity} = 100\%\ \text{transmission} - \%\ \text{actually transmitted} \qquad (12\text{-}1)$$

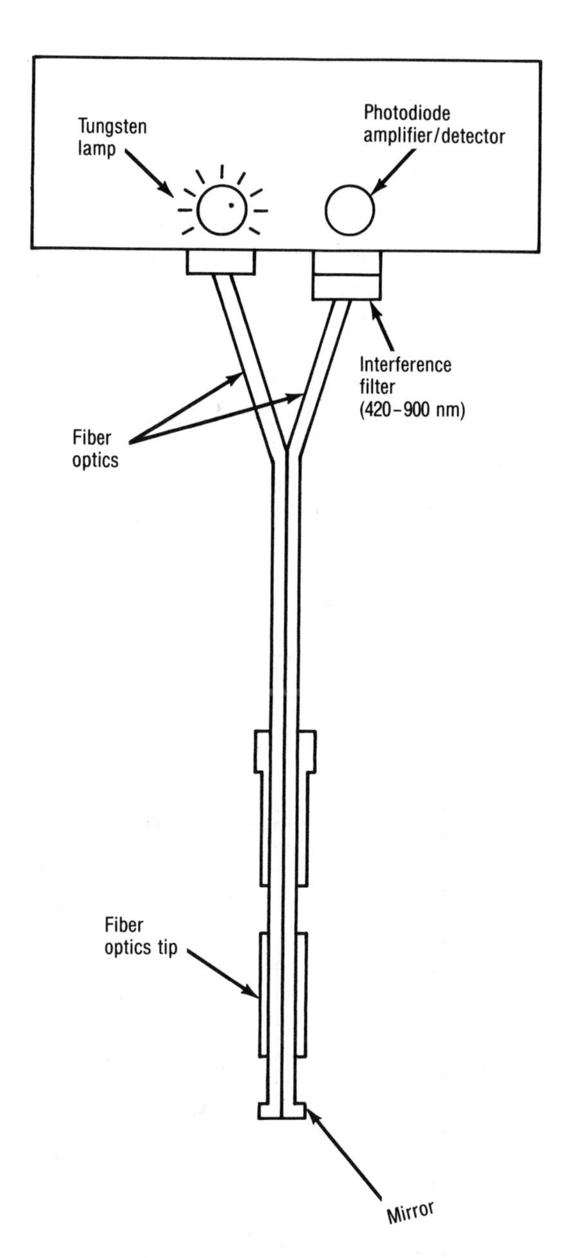

Figure 12-9
Operational Schematic Diagram for Fiber Optic Probe Colorimeter

(reprinted by permission from SPIE)

A measure of the ability of airborne particulate to attenuate light is the optical density defined by:

$$\text{Optical Density} = \log_{10} \frac{1}{1 - \text{Opacity}} \tag{12-2}$$

Figure 12-10 shows a transmissive sensor used to characterize smoke in a stack. The opacity is correlated with the efficiency of the manufacturing process. Generally, an increase in opacity indicates that the process being monitored is running at less than optimum conditions.

The scattering concept can be expanded to determine the volume fraction of one immiscible liquid in another.[12] Figure 12-11 shows a sensor designed for detecting oil droplets in water. The oil-water mixture flows into a chamber where a light beam is projected perpendicular to the path of the flow. The ratio of the light entering the off-axis probe to the light entering the on-axis probe is a measure of the oil concentration.

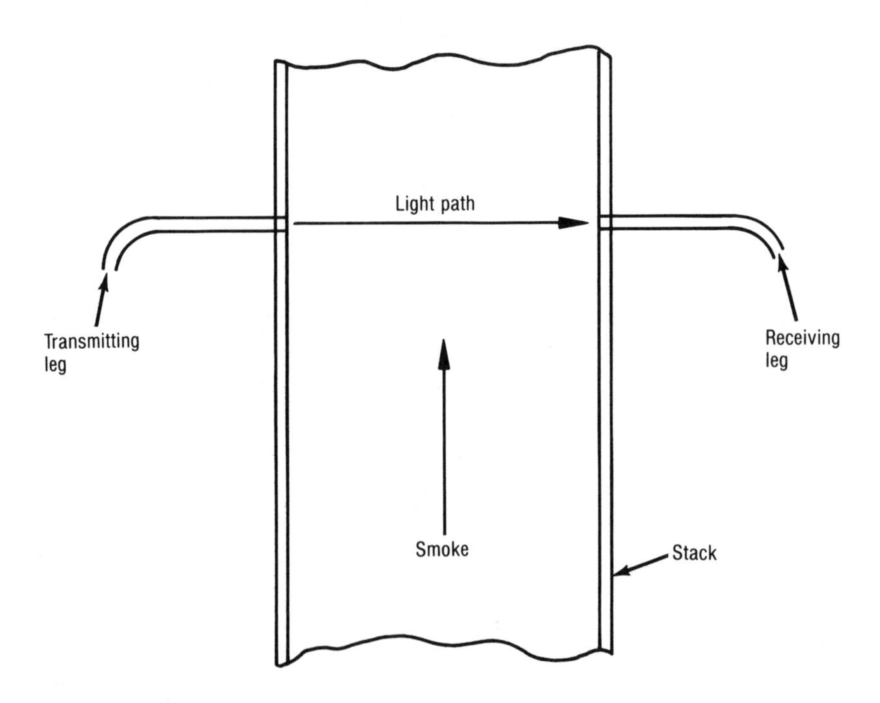

Figure 12-10
Transmissive Sensor for Smoke Characterization in Stack

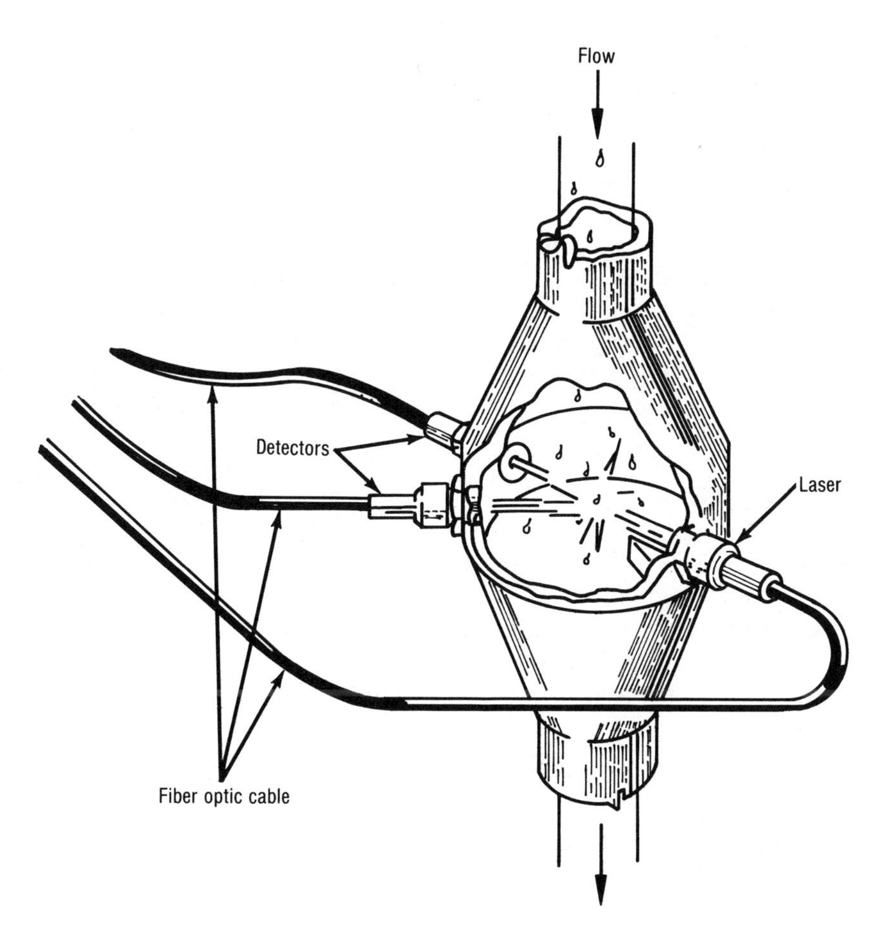

Figure 12-11
Transmissive Sensor for the Detection of Oil in Water Using Scattering

Scattering has also been used to detect physical transitions such as melting, the glass transition, and crystallization.(18) If the material to be characterized is transparent in one state, the change in state generally results in the formation of scattering sites, which denotes the physical change. Since the physical change of state usually occurs at a specific temperature, the scattering phenomena can be used as the basis for a temperature switch. Devices that use the concept have a pellet of material in a reflective container. In the transparent state, the reflection is detected from the container. In the high scattering state, the reflectivity is greatly diminished.

Scattering has been used in a transmissive system to fabricate dosimeters that detect radiation level. In the presence of nuclear radiation, many glasses form color centers that act as scattering sites. The transmission level in a radiation-sensitive glass fiber is a measure of radiation exposure.

Molecular constituents can be identified by their Raman-scattered spectra. Raman laser scattering has been used to detect small concentrations of various gases.[(19)] The system employs a high intensity monochromatic light source coupled to the transmitting fiber that illuminates the gas sample. The receiving fiber collects the Raman-scattered radiation that is characteristic of the particular gas being measured.

Applications of this technique include monitoring methane vapor, which can, in turn, identify a possible liquid natural gas spill. Detection of 1% or less of methane vapor is achievable.

To increase the overall system sensitivity, a multipass sampling cell was used, as shown in Figure 12-12.[(19)] The spherical mirrors labeled M_1 and M_1' provide for multiple passes of the light beam. A portion of the scattered light is injected directly into the receiving fiber, while additional scattered light reflects from mirror M_2 into the receiving fiber. The fiber chosen for this sensor was glass, not plastic, due to the fact that the fibers themselves have a characteristic Raman-scattering spectrum. Plastic fiber has many intensity peaks that mask the gas spectra to be measured.

REFRACTIVE INDEX CHANGE

The output level and the pattern of light are related to the index of refraction difference between the core and cladding. Figure 12-13 shows the effects

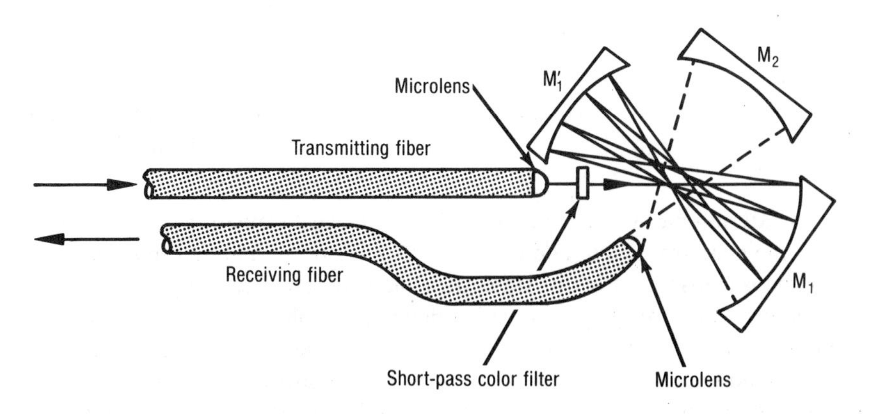

Figure 12-12
Multipass Reflection Sensor Configuration for Raman Scattering

of the three principal indices of refraction conditions and the resulting ray paths.[20] Sensors using this concept have been devised either by having the substance (gas or liquid) in contact directly on an unclad glass rod (or prism) that acts as the core or by reacting with a permeable coating on the glass rod. Gases such as NH_3, HCN, and H_2S have been detected by this approach.[21]

Figure 12-14 shows a reflective fiber optic probe coupled to a prism. In air, the index of the prism is sufficiently higher than that of air, and a condition of total internal reflection exists within the prism. Therefore, light transmitted to the prism is reflected efficiently back into the fiber optic

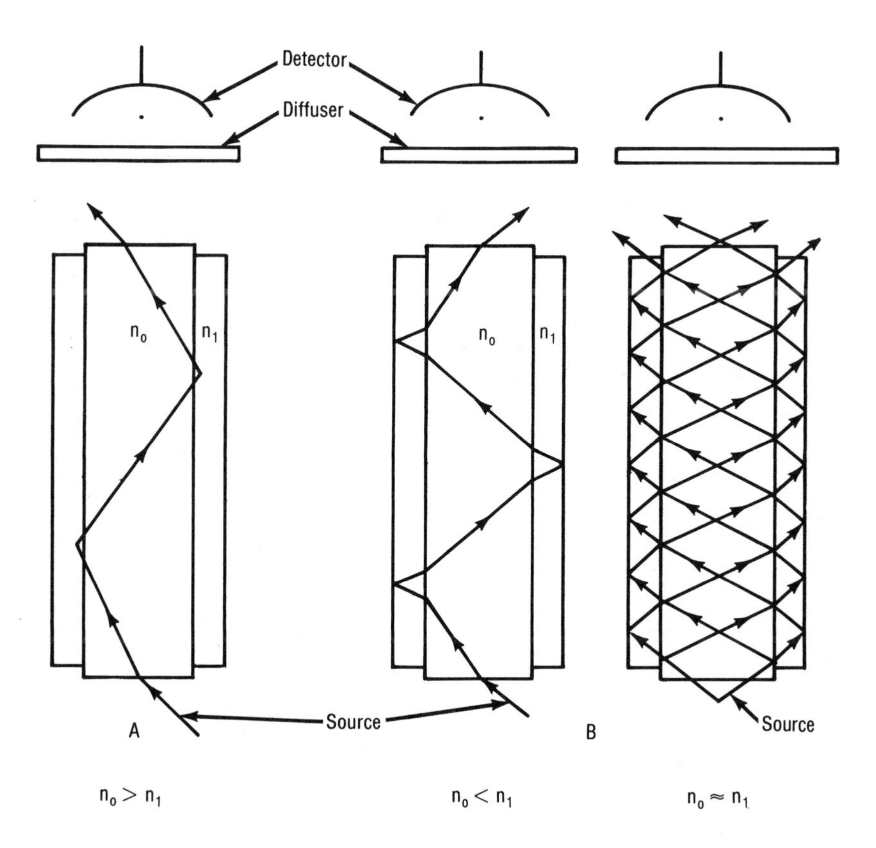

Figure 12-13
Ray Paths Associated with Various Index of Refraction Conditions in the Cladding
(patented)

probe. However, if the prism is immersed in a liquid with a refractive index approaching the prism value, the reflected light significantly decreases. If the index of the liquid is higher than the prism, the intensity will drop to even lower values, typically less than 10% of the initial value.

While the concept is normally used for liquid level switch-type sensors, it can provide an analog signal proportional to the refractive index of the liquid. Since many components of a solution alter the refractive index of the liquid, as a function of concentration, such a device can provide a measure of concentration. An example would be the concentration of sulfuric acid, which has a refractive index of just above 1.33 in low concentrations and 1.46 for a 90% level. The device can also be used to differentiate liquids of different refractive indices, such as gasoline and water.

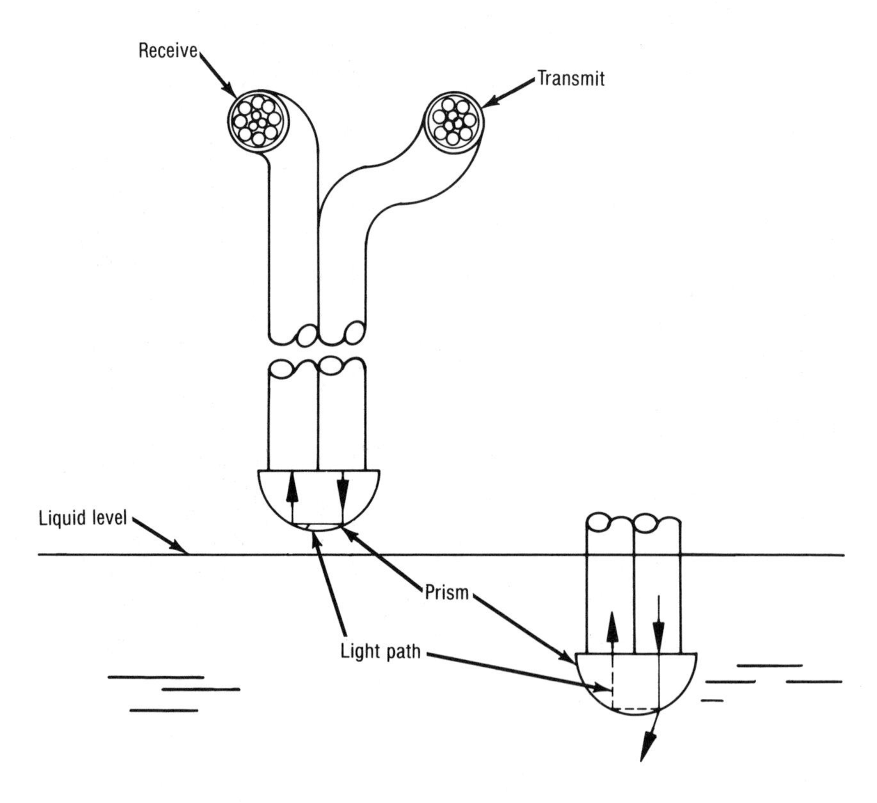

Figure 12-14
Refractive Index Change Sensor Using a Reflective Fiber Optic Probe

The prism approach has also been used to measure void fraction.[21] As a bubble or void passes by the prism in a volume of material, the liquid and the voids have different refractive indices, which provides a signal corresponding to which condition is at the probe tip. The probe, in effect, is used as a counter per unit time that determines the void fraction. The accuracy is estimated to be 0.5%.

INTERFEROMETRY

Nearly all the chemical sensors use intensity-modulation schemes. However, some chemical reagents swell when the reaction occurs. The potential of a mechanical (axial) movement corresponding to the concentration of a chemical species allows interferometric sensors to be considered. The principle is identical to electric and magnetic field interferometric sensors using piezoelectric and magnetostrictive coatings. Butler[25] has reported coating a fiber with palladium. The palladium is a hydrogen absorber that expands with hydrogen concentration. The palladium-coated fiber is the sensing leg in a Mach-Zehnder interferometer.

APPLICATIONS

The applications for chemical analysis fall primarily into two categories: industrial process monitoring and medical monitoring. The primary advantage for both categories is *in situ* real-time measurement with minimal perturbations to the process being monitored.

Remote chemical analysis has a distinct advantage in a chemical plant. Centralized analyzers using remote fiber optic sensors can potentially replace costly in-line conventional analysis equipment. The evolution of the sensor technology will be slowed by the fact that most materials do not fluoresce or have absorption or scattering phenomena in the desired wavelength range. Therefore, reagents have to be developed that are compatible with optical techniques for the vast majority of materials and processes to be monitored.

The medical sensor area is of keen interest because of the potential of extremely small-size sensors for direct use within the body. Fluorescent techniques are much more widely applicable in biological processes. For instance, fluorescence can differentiate between diseased and normal arterial tissue as well as between many cancerous and noncancerous tissues.

REFERENCES

1. Chabay, I., 1982, "Optical Waveguides," *Analytical Chemistry*, Vol. 54, No. 9, pp 1071A–1080A.

2. Graff, G., February 1983, "Fiber Optics Analyze Chemical Processes," *High Technology*, pp 24–25.

3. Maugh, T. H., II, November 1982, "Remote Spectrometry with Fiber Optics," *Science*, Vol. 218, pp 875–876.

4. Grant, W. B., October 1983, "Laser Spectroscopy Techniques Make Industrial Appearance," *Industrial Research and Development*, pp 154–57.

5. Anonymous, 1982, "Fiber Optics Simplify Remote Analyses," *C&EN*, pp 28–30.

6. Chudyk, W., Kenhy, K., Jarvis, G., and Pohlig, K., 1986, "Monitoring of Ground Water Contaminants Using Laser Fluorescence and Fiber Optics," *Proceedings of the ISA*, Houston, TX, pp 1237–1243.

7. Sarri, L., and Weitz, W., 1983, "Immobilized Morin as Fluorescence Sensor for Determination of Aluminum (III)," *Analytical Chemistry*, Vol. 55, pp 667–670.

8. Peterson, J., Fitzgerald R., and Buckhold, D., 1984, "Fiber Optic Probe for *in vivo* Measurement of Oxygen Partial Pressure," *Analytical Chemistry*, Vol. 56, pp 62–67.

9. Seitz, W., 1984, "Chemical Sensors Based on Fiber Optics," *Analytical Chemistry*, Vol. 56, No 1, pp 19A–34A.

10. Urbano, E., Offenbacher, H., and Wolfbeis, O., 1984, "Optical Sensor for Continuous Determination of Halides," *Analytical Chemistry*, Vol. 56, No. 3, pp 427–429.

11. Sarri, L., and Seitz, W., 1982, "pH Sensor Based on Immobilized Fluoresceinamine," *Analytical Chemistry*, Vol. 54, pp 821–823.

12. Chamberlin, R., and Nellist, J., 1985, "The Applications of Fiber Optics in Worldwide Offshore Oil and Gas Industry Operations," Distributed by Gulf Fibercom, Inc., Irvine, CA.

13. Schirmer, R., 1986, "On-Line, Fiber Optic-Based Near Infrared Absorption Spectrophotometry for Process Control," *Proceedings of the ISA*, Houston, TX, pp 1229–1235.

14. Murry, R., Smith, D., and Wright, P., 1986, "Fiber-Optic Sensors for the Chemical Industry," *Optics News*, pp 31–33.

15. Seitz, W., August 1985, "Chemical Sensors Based on Fiber Optics," *Sensors*, pp 7–9.

16. Giuliani, J., Wohltjen, W., and Jarvis, N., 1983, "Reversible Optical Waveguide Sensor for Ammonia Vapors," *Optical Letters*, Vol. 8, No. 1, pp 54–56.

17. Anonymous, 1982, "Smoke Opacity Monitoring Systems," *Applications Engineering Handbook*, Dynatron, Wallingford, CT.

18. Frushour, B., Sabatelli, D., and Fisher W., 1986, "Fiber Optic Probe for Simultaneous Measurement of Light Transmission and Heat Flow in a Differential Scanning Colorimeter," To Be Published.

19. Chang, R., and Benner, R., 1979, "Laser-Raman Point Monitoring of CH4 Vapor in the LNG Storage Field," Gas Research Institute, Research Grant No. 5014-363-0146.

20. Handy, E., David, D., Kapany, N., and Unterleitner, F., 1975, "Coated Optical Guides for Spectrophotometry of Chemical Reactions," *Nature*, Vol. 257, pp 646–647.

21. Handy, E., and David, D., 1977, "Optical Analytical Device, Waveguide and Method," U.S. Patent 4,050,895.

22. Graindorge, P., LeBoudex, G., Meyet, D., and Arditty, H., 1985, "High Bandwidth Two-Phase Flow Void Fraction Fiber Optic Sensor," *Proceedings SPIE-Fiber Optic Sensors*, Vol. 586, pp 211–214.

23. Munsinger, R., 1981, "Fiber Optic Colorimetry," *Electro-Optical Systems Design.*

24. Hammel, M., February 1981, "Using Fiber Optics in Colorimetry," *Modern Paint and Coatings*, pp 53–56.

25. Butler, M., September 1984, "Chemical Sensor Uses Optical Fibers," *C&EN*, pp 22–23.

Rotation Rate Sensors (Gyroscopes)

13

INTRODUCTION

The major advantages of a fiber optic gyroscope over mechanical devices include:[1,2] no moving parts, no warm-up time, unlimited shelf life, minimal maintenance, large dynamic range, and small size. As the technology advances, low cost sensors are projected.

The scope of applications is quite broad with a wide range of specifications. Figure 13-1 graphically shows the various applications in relation to the required dynamic range and sensitivity.[3]

All optical rotation sensors are based on the Sagnac effect, which is described in Chapter 4.

SENSOR MECHANISM

The Sagnac effect uses an interferometric technique for rotation rate detection.[4] The initial light beam is split into two beams that travel along a single fiber in a coiled configuration, as shown in Figure 13-2. One path is clockwise, and the other is counterclockwise. When the fiber ring rotates in a clockwise direction, the light propagation in the clockwise direction is longer. This situation is due to the fact that the starting point has now moved due to rotation, and the light beam must travel a greater distance to reach the starting point. Conversely, the counterclockwise light beam travels a shorter distance. The path length difference results in a phase difference that affects the output of the Sagnac interferometer in a manner that is related to the rotation rate.

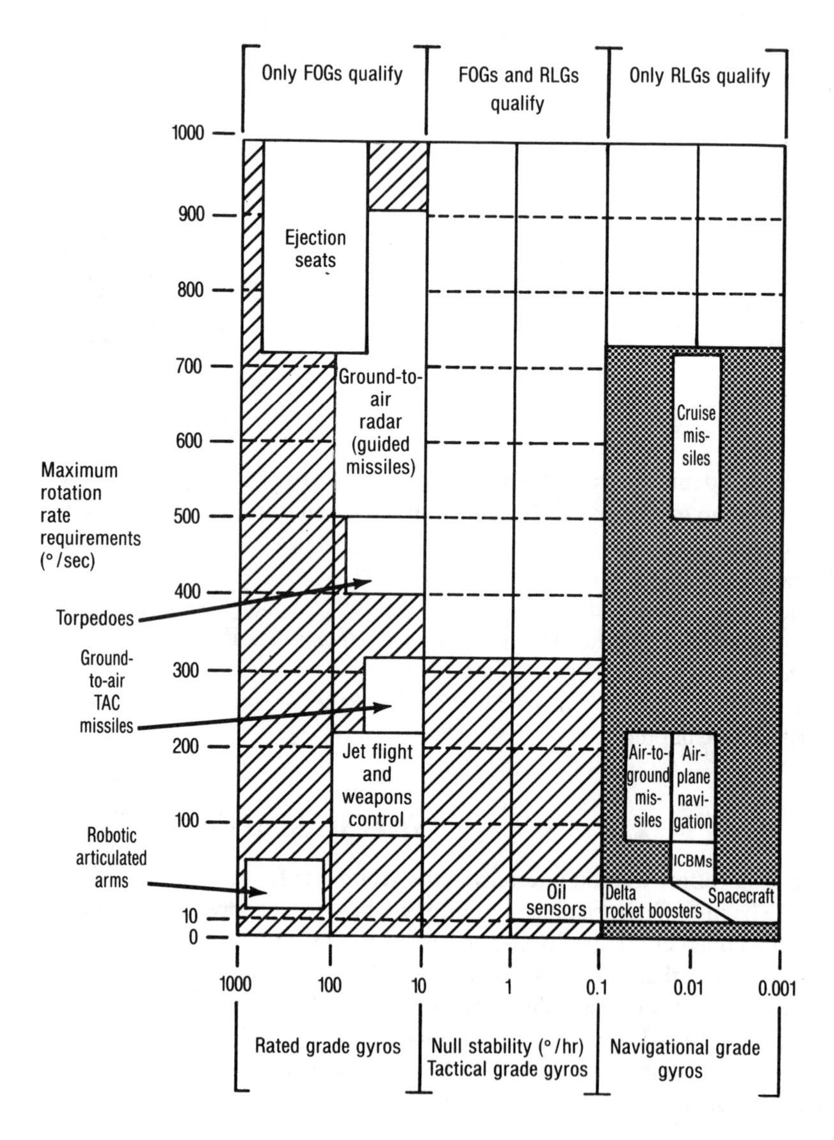

Figure 13-1
Applications of Optical Gyroscopes with the Required Specifications[3]

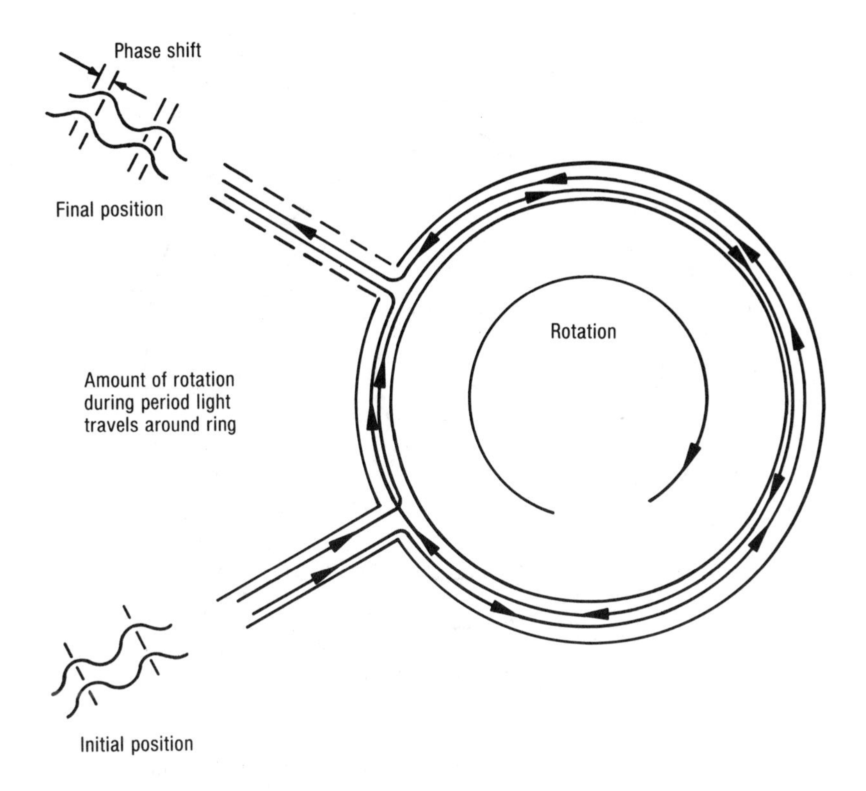

Figure 13-2
Sagnac Effect in a Coiled Fiber Used for Rotation Rate Sensing

Figure 13-3 provides the definition for the mathematical relationship between the path length difference, ΔL, and the rotation rate Ω, which is given by:[(2)]

$$\Delta L = (4A/c)\,\Omega \qquad (13\text{-}1)$$

where A is the area enclosed by the light path (of radius R) and c is the velocity of light in a vacuum. For a clockwise rotation, starting at point 1 and moving to point 2, the clockwise path length is $2\pi R + \Delta S$. The counter-clockwise path length is $2\pi R - \Delta S$. ΔS is the change in circumferential path length. Therefore:

$$\Delta L = 2\pi R + \Delta S - [(2\pi R - \Delta S)] = 2\Delta S \qquad (13\text{-}2)$$

The path length traveled by the light in a time Δt, is given by:

$$\Delta L = c\Delta t = (4A/c)\,\Omega \qquad (13\text{-}3)$$

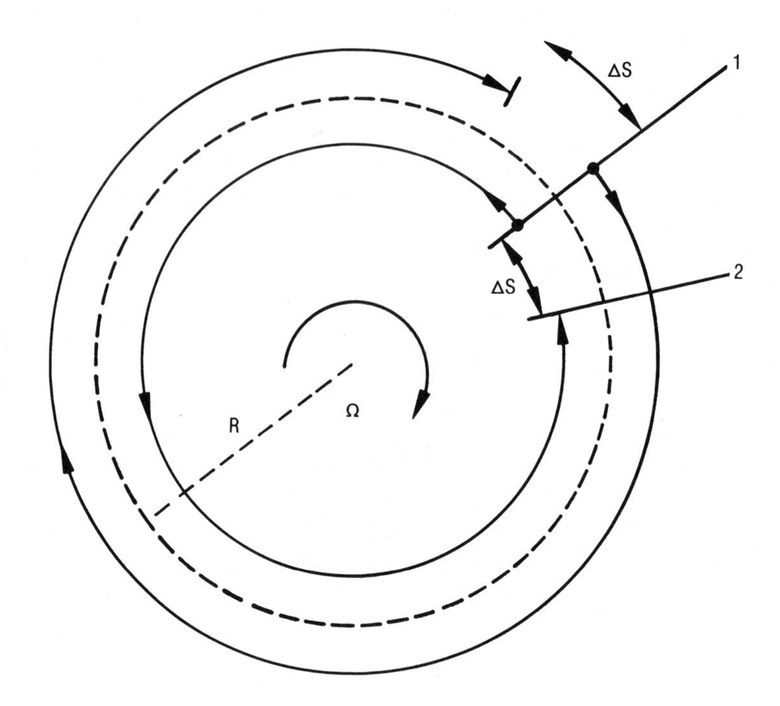

Figure 13-3
Sagnac Configuration with Geometric Parameters Defined

(reprinted by permission from Dynamic Systems, Inc.)

Both the definition of ΔL and Δt defined above for a vacuum are applicable in a medium of refractive index, n.[(2)] For a fiber coil with N turns, Δt then becomes:

$$\Delta t = (4AN/c^2)\,\Omega \tag{13-4}$$

A phase shift of $\Delta\phi$ is defined as:

$$\Delta\phi = 2\,n\Delta t\,\frac{c}{\lambda_0} \tag{13-5}$$

Since $\lambda_0 = \lambda n$ (λ_0 is the wavelength in vacuum),

$$\begin{aligned}\Delta\phi &= 2\,\pi\Delta t\,\frac{c}{\lambda n} \\ &= \frac{(8\,\pi\,AN)}{(\lambda_0\,c)}\,\Omega\end{aligned} \tag{13-6}$$

And ΔL is defined by:

$$\Delta L = \left[\frac{4\,AN}{c}\right]\Omega \tag{13-7}$$

For a fiber with a coil diameter D and length L the following equations apply:

$$A = \pi D^2/4 \tag{13-8}$$

$$N = L/\pi D \tag{13-9}$$

Therefore:

$$\Delta\phi = (2\,\pi\,LD/\lambda_0\,c)\,\Omega \tag{13-10}$$

$$\Delta L = (LD/c)\,\Omega \tag{13-11}$$

Using typical values in Equation 13-11, the magnitude of ΔL is on the order of 10^{-15} cm for a single fiber loop.(2) To increase ΔL to a more reasonable value, the number of turns in the fiber loop needs to be quite large.

As discussed in Chapter 4, the maximum sensitivity occurs when the phase changes are $\pi/2$ or some multiple thereof. Therefore, to maximize the sensitivity for small rotations, a nonreciprocal, stable $\pi/2$ phase shift must be introduced between the counter rotating beams.(15) Figure 13-4 shows two biasing approaches. The approach applies a bias so that at small rotation rates, the inherent insensitivity of the sensor, due to the output being near a maximum point (little slope change), is shifted to a maximum slope change point. A more common approach is to sinusoidally modulate the phase difference between counterpropagating light beams. The ac detector current versus rotation rate is shown in Figure 13-4(b). The sensing system now has a maximum sensitivity at zero rotation. The response is linear over a limited region of the response curve and direction can be determined.

The fiber optic gyroscope using a modulator is shown in Figure 13-5.(5) The spatial filter and polarizer ensure that the light paths for the counter-propagating waves are the same. The phase modulator is typically a fiber wrapped on a piezoelectric cylinder. Through a feedback circuit, the piezoelectric cylinder and, hence, fiber length are altered in a sinusoidal manner to achieve the desired nonreciprocal phase shift, which, in turn, provides the bias. Those using this configuration are referred to as analog gyroscopes. They are limited to applications with moderate dynamic range, sensitivity, and drift. The phase shift and length change are defined by Equations 13-6 and 13-7, respectively.

Digital fiber optic gyroscopes are used for applications that require high dynamic range, sensitivity, and low drift. The sensor configuration is shown in Figure 13-6.[5] In contrast to the analog configuration, the digital approach adds two frequency shifters that introduce a frequency difference, Δf, between the counterpropagating beams. The path length difference, ΔL, results in a frequency difference, Δf, as defined below:

$$\Delta L = \frac{\Delta f L n \lambda_0}{c} \tag{13-12}$$

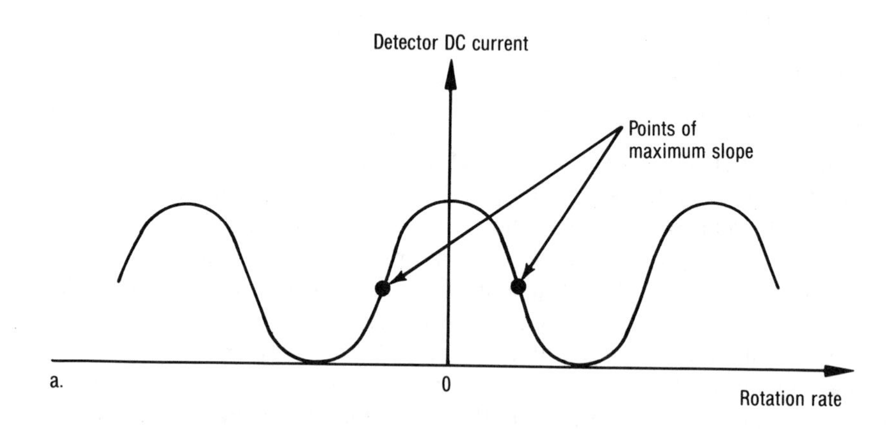

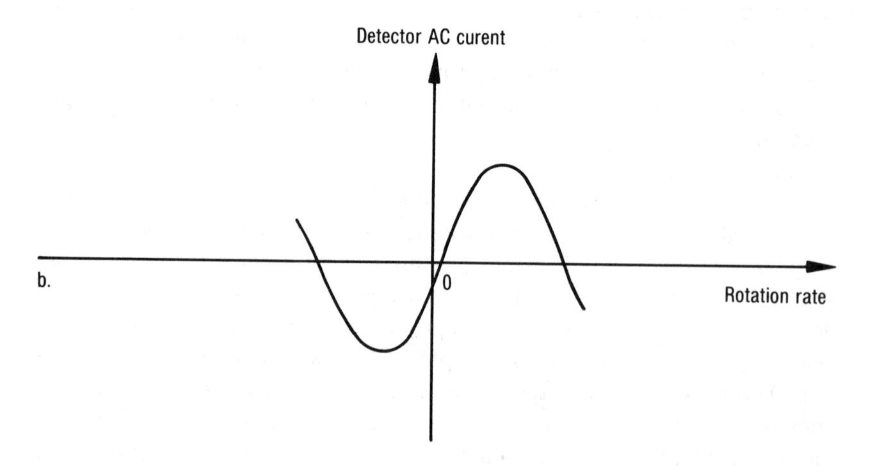

Figure 13-4
Biasing Approaches

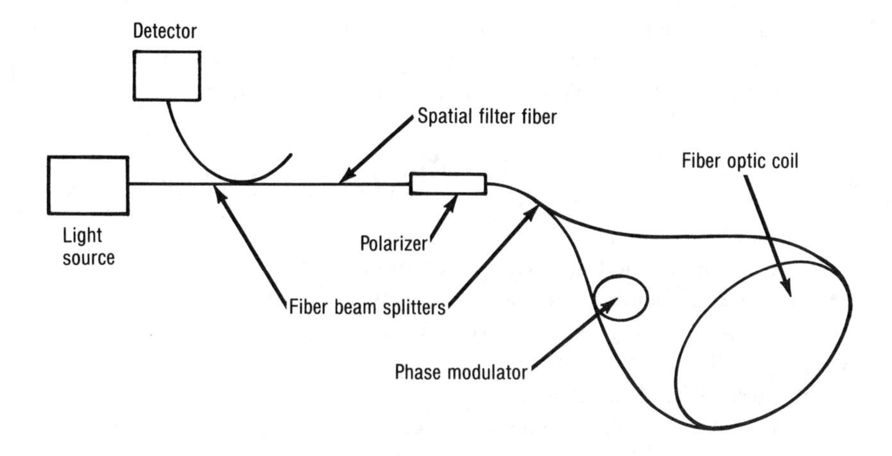

Figure 13-5
Analog Fiber Optic Gyroscope Configuration

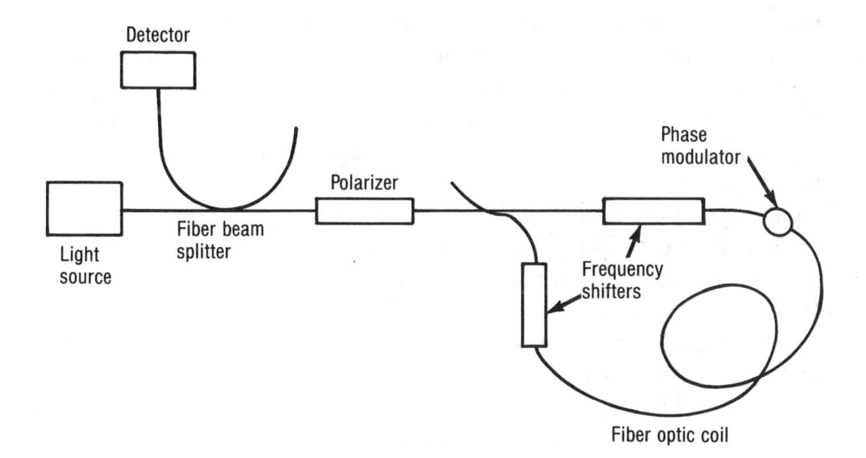

Figure 13-6
Digital Fiber Optic Gyroscope Configuration

For the system to be at null with maximum sensitivity, the change of a path length due to rotation must be equal to the path difference associated with the frequency shift. Therefore:

$$\Delta f = \frac{4\ NA\ \Omega}{\lambda_0 nL} \tag{13-13}$$

The analog fiber optic gyroscope has a dynamic range of 10^3 to 10^5 with a scale fraction correction (sensor accuracy) of 0.3 to 3%. The digital gyroscope has a dynamic range of 10^8 and a scale factor correction of 0.03%.[5]

NOISE LIMITATIONS

Several sources of noise may be present in a fiber optic gyroscope:[2] Rayleigh scattering in the fiber, scattering from interfaces, polarization effects, temperature gradients, stress-induced effects, and magnetic field effects.

Rayleigh scattering and scattering from interfaces have the same basic effect.[6,7,8] The scattered light, if it is at least partially coherent with the transmitted light, can interfere and cause noise. Light sources with long coherence lengths, i.e., sources in which the light beam wavefront stays coherent for long distances, create a severe backscatter noise problem. Light sources that have shorter coherence lengths (more diffuse) reduce this noise problem.

Single-mode fiber is used in nearly all such gyroscopes. However, polarization effects can cause a degeneration into two modes.[9,10,11] Since the two polarization modes have different propagation constants, the modes interfere with the resultant noise, degrading sensor performance. To overcome this problem, a polarizing lens is used to inject a given polarization state in the fiber. Polarization maintaining fiber is used to preserve the polarization state in the fiber.

Other parameters, such as temperature gradients, induced stresses, and magnetic field effects (Faraday rotation), alter the polarization state. High birefringence polarization-maintaining fiber tends to minimize these noise effects.

In summary, fiber optic gyroscopes have distinct advantages over mechanical and other optic gyroscopes. Sensitivity ranges have been reported for $1°/\text{hr}$ to $0.01°/\text{hr}$.[5,6,12] The ultimate sensitivity, based on realistic fiber properties, is calculated to be $3 \times 10^{-4}°/\text{hr}$.[10] Polarization-maintaining fiber is required to minimize sensor noise problems.

REFERENCES

1. Kim, B., and Shaw, H., March 1986, "Fiber-Optic Gyroscopes," *IEEE Spectrum*, pp 54–60.

2. Davis, C. M., et al. 1982, *Fiber Optic Sensor Technology Handbook*, Dynamic Systems, Reston, Virginia.

3. Kreidl, J., January 1987, "Northrop's New RLG to Rival the FOG," *Lightwave*, p. 58.

4. Hecht, J., July/August 1982, "Fiber Optics Turns to Sensing," *High Technology*, pp 49-56.

5. Udd, E., December 1985, "Fiber Optic vs. Ring Laser Gyros: An Accessment of the Technology," *Laser Focus/Electro-optics*, pp 64-74.

6. Giallorenze, T., Bucaro, J., Dandridge, A., Sigel, G. H., Cole, J., Rashleigh, S., and Prest, R., 1982, "Optical Fiber Sensor Technology," *IEEE Journal of Quantum Electronics*, Vol. QE-18, No. 4, pp 626-664.

7. Giles, I., McMillan, J., Mackintosh, J., and Culshaw, B., 1985, "Coherence in Optical Fibre Gyroscopes," *SPIE-Fiber Optic Sensors*, Vol. 586, pp 180-186.

8. Bohn, K., Marten, P., Petermann, K., and Weidel, E., 1981, "Low Drift Fibre Gyro Using Superluminescent Diode," *Electronics Letters*, Vol. 17, No. 10, pp 352-353.

9. Burns, W., Moeller, R., Villarruel, C., and Abebe, M., 1983, "Fiber-Optic Gyroscope with Polarization-Holding Fiber," *Optics Letters*, Vol. 8, No. 10, pp 540-542.

10. Jeunhomme, L., 1983, *Single-Mode Fiber Optics*, Marcel Dekker, New York, pp 251-269.

11. Burns, W., Chen, C., and Moeller, R., 1983, "Fiber-Optic Gyroscopes with Broad Band Sources," *Journal of Lightwave Technology*, Vol. LT-1, No. 1, pp 78-104.

12. Dianov, E., et al., 1985, "Rotation Sensors Based on Single-Mode Fibers with Low and High Birefringence," Third International Conference in Optical Fiber Sensors, *Technical Digest*, San Diego, CA, p. 120.

Distributive Sensing Systems

14

INTRODUCTION

A distributive fiber optic sensing system is defined as a network of sensors using a common data highway to communicate back to a central area. The elements of an all-fiber optic network include a centralized common light source; optical fibers to carry light to optically passive sensors, and a return fiber path to carry modulated light to a photodetector.(1) An all-fiber optical system has the advantages of electrical isolation and noise immunity for both the sensing and communication system. Hybrid systems are also used in which passive optical fibers and sensors work in conjunction with active electro-optic conversion nodes.(3)

All the sensing concepts can be used in distributive networks, including intensity-modulated and phase-modulated sensors. Sensor concepts such as reflection and interferometry require that the sensor be inserted in the system through couplers. Couplers introduce loss, so the number of such sensors in a network is limited. Intrinsic-type sensors that are incorporated in the fiber itself lend themselves easily to distributive systems because couplers are not required and the fiber is used in a continuous link. Some examples of such sensor types include chemically doped fibers (core or cladding), which change transmission level due to the local environmental changes, and microbending, in which local perturbations alter fiber transmission. Figure 14-1 shows examples of intrinsic sensors used in a continuous single fiber network. Other multiple sensor configurations include star-coupled and "T"-coupled networks.

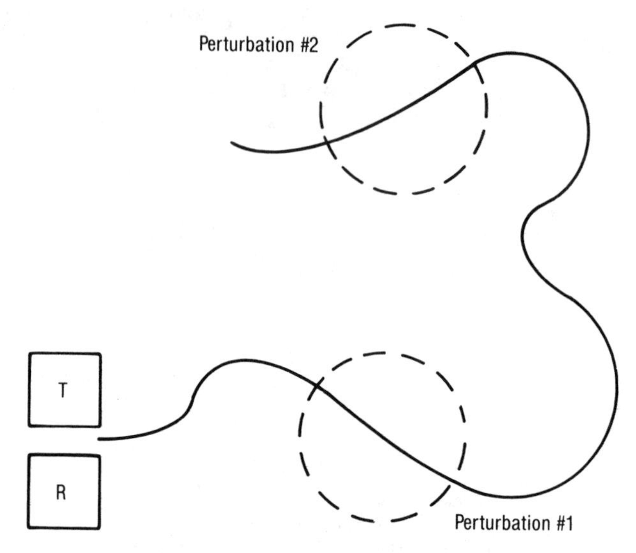

(a) Chemical Doped Fiber which Changes Transmission with Temperature or Some Other Physical Change

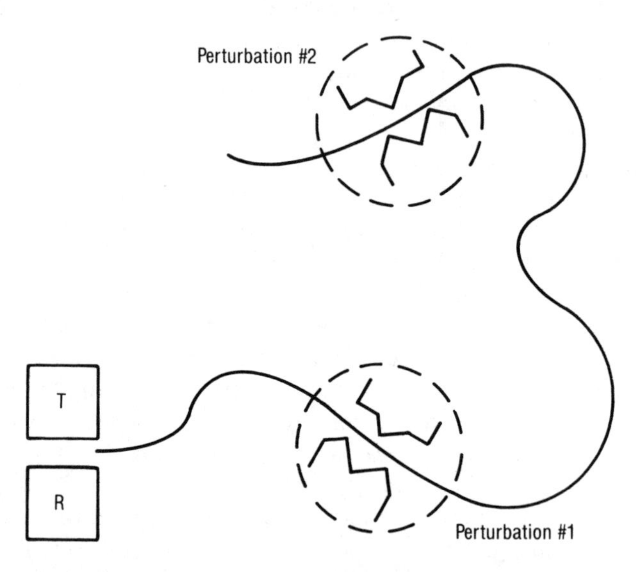

(b) Microbending Fiber in which Fixture Squeezes Fiber as a Function of Temperature or Some Other Physical Change

Figure 14-1
Intrinsic Sensors in a Continuous Single Fiber Sensing Network

As the sensors are configured in Figure 14-1, it is possible to determine that a perturbation has occurred and its magnitude, only if one sensing point is activated. If multiple points are perturbed, the information is masked. Only the fact that a perturbation occurred somewhere on the fiber can be determined. To identify each individual sensor, wavelength division or time division multiplexing must be used.[2] In a wavelength division multiplexing scheme, each sensor corresponds to a specified wavelength band. In a time division multiplexing scheme, each sensor corresponds to a different time delay slot. Wavelength division multiplexing systems use narrow band filters or sensing materials to generate light in a narrow wavelength band. Time division multiplexing uses a form of optical radar called optical time domain reflectometry (OTDR).[4] The concept is described later in this chapter.

HYBRID SYSTEMS

Hybrid systems use fiber optic sensors, optical fibers for transmission, and electro-optic nodes. Figure 14-2 depicts such a system. The fiber optic sensors are generally intensity-modulated and can be either digital switches or analog devices. The electro-optic node provides the light source and the detection electronics. The detected optical signal is converted into an optoelectronic signal, then back into an electro-optical signal and is transmitted on the fiber optic highway to the main control area. Since the electro-optic node acts as a repeater, there is no real limit to the number of sensors that can function in the system. However, the node does require power and shielding to prevent unacceptable environmental noise.

MULTIPLEXING SCHEMES

Wavelength division multiplexing requires that each sensor in the network correspond to a specific wavelength band.[1] The light source can be an LED transmitting at a specific wavelength band. However, such a configuration requires that each wavelength have its own light source. A preferred system would use a single broadband (white) light source with a filter of the desired wavelength injected into the system just prior to each sensor, as shown in Figure 14-3. The sensors typically would be intensity-modulated units since interferometric sensors in this type of network would require multiple laser sources. The sensor-modulated information is now wavelength encoded. The information is carried back over a common fiber and then separated by filters of the specified wavelength into separate photodetectors for each channel. Typically, each sensing channel will require a spectral width of 20 to

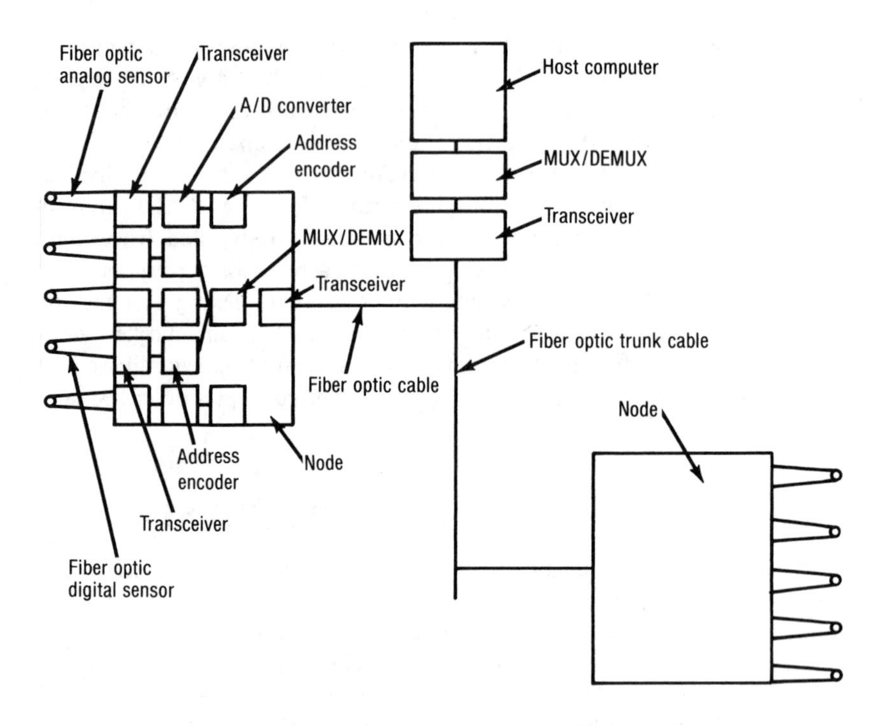

Figure 14-2
Hybrid System Using Passive Fiber Optic Components and Electro-optic Node

50 nm to achieve minimal interference from one channel to another. The spectral width characteristic limits the network capacity to about 20 sensors. The requirement for sensors and filters to be coupled into the network also limits the number of possible sensors.

Time division multiplexing is generally a simplier system to facilitate. In addition, such a system can be used with intensity or phase-modulated sensors. A basic approach to the time division multiplexing scheme uses optical time domain reflectometry (OTDR). A pulse is sent out along a single fiber. The return signal associated with Rayleigh scattering is dependent upon fiber attenuation along its length (see Figure 14-4).[4] The backscatter is used to identify sensor positions. Figure 14-5[4] shows a plot of backscatter amplitude versus range (delay time). The slope of the curve defines fiber attenuation. The attenuation changes associated with various sensors are indicated. The incremental loss at each sensor gives a measure of the magnitude of the parameter being monitored at that position. Optical

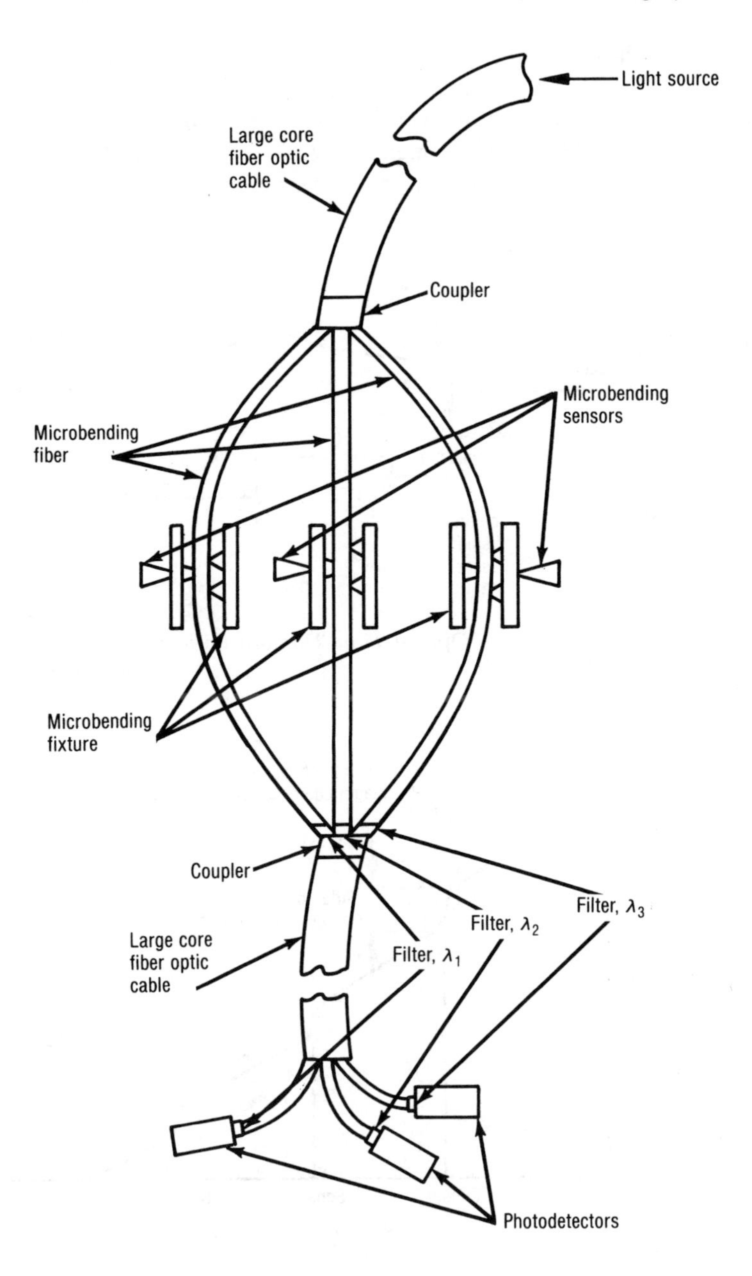

Figure 14-3
Wavelength Division Multiplexing Scheme

time domain reflectometry has a drawback in that it cannot resolve very short distances from the light source along the optical fiber without complicated and expensive high speed electronics. This problem can be resolved by using the optical frequency domain reflectometry (OFDR). In such a detection system, the return signal is mixed with a reference signal that allows short distances to be resolved.[4,5]

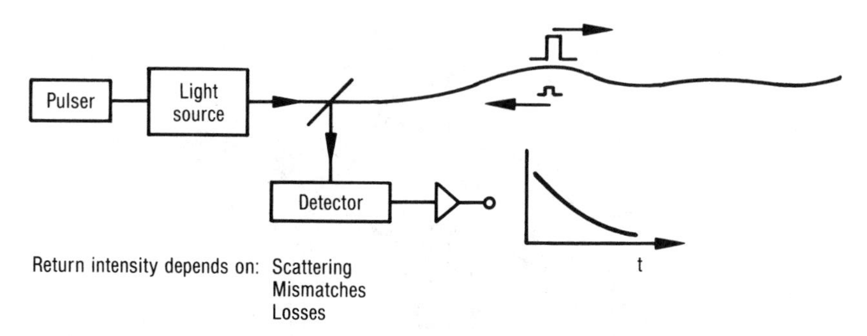

Figure 14-4
Optical Time Domain Reflectometry (OTDR) with Single Optical Fiber

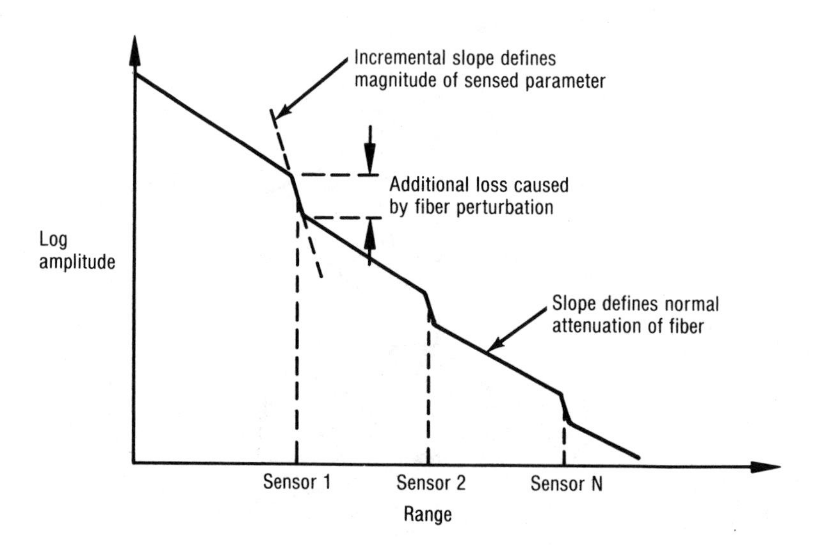

Figure 14-5
OTDR Characteristic for a Distributed Intrinsic Fiber Optic Sensing System

SENSOR NETWORK CONFIGURATION

The simplest configuration of a distributed sensing system is a single continuous fiber in which the sensing function is either incorporated intrinsically in the fiber, or a microbending device is attached at various sensing locations. Typically, this approach uses an OTDR to determine optical reflection discontinuities, as shown in Figure 14-6.[2] The activation of any sensor in the network must still allow sufficient light to pass so that subsequent in-line sensors can function.

A "T"-coupled network is shown in Figure 14-7. The sensors use the reflective concept and transmit and receive on a single fiber so that only one coupler into the main fiber is required. The couplers in the system will have

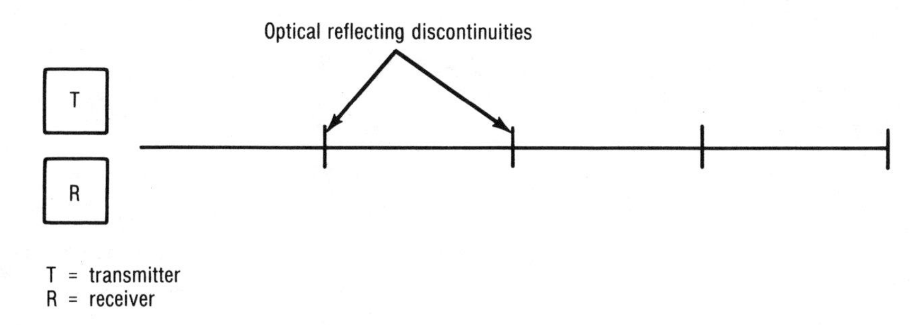

Figure 14-6
Single Continuous Fiber Network

(reprinted by permission of Dynamic Systems, Inc.)

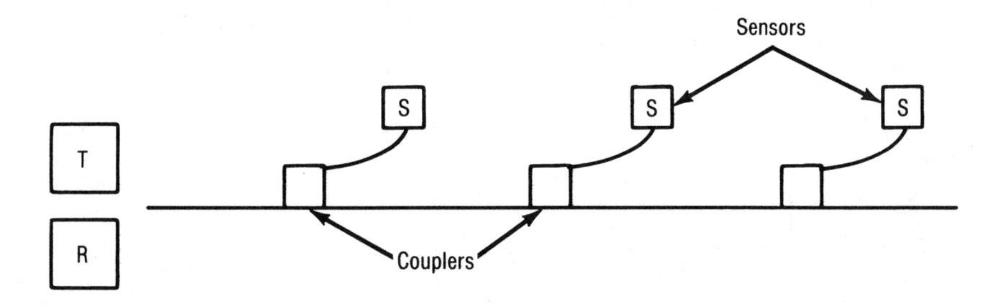

Figure 14-7
"T" Coupled Network (Reflective Sensors)

(reprinted by permission from SPIE)

some backscatter loss associated with them. The sensor reflective light intensity level is superimposed upon the connector signal. Figure 14-8 shows a "T"-coupled network in which a series of interferometers are monitored. As discussed previously, the phase shift associated with an interferometer causes both destructive and constructive interference, which changes the returning light intensity. The backscatter light level is, therefore, different at the entrance coupler and the exit coupler, and the sensor is monitored.

The "T"-coupled network can easily be expanded to a star configuration. Figure 14-9 depicts such a configuration.[4] This network approach requires that each leg of the star be of a different length to allow a different time

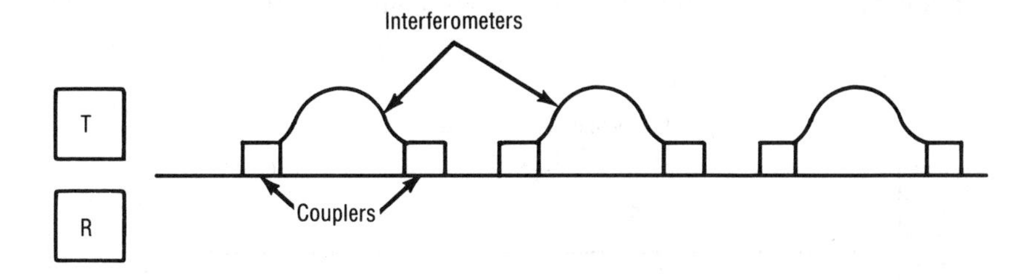

Figure 14-8
"T" Coupled Network (Interferometer Sensors)
(reprinted by permission from SPIE)

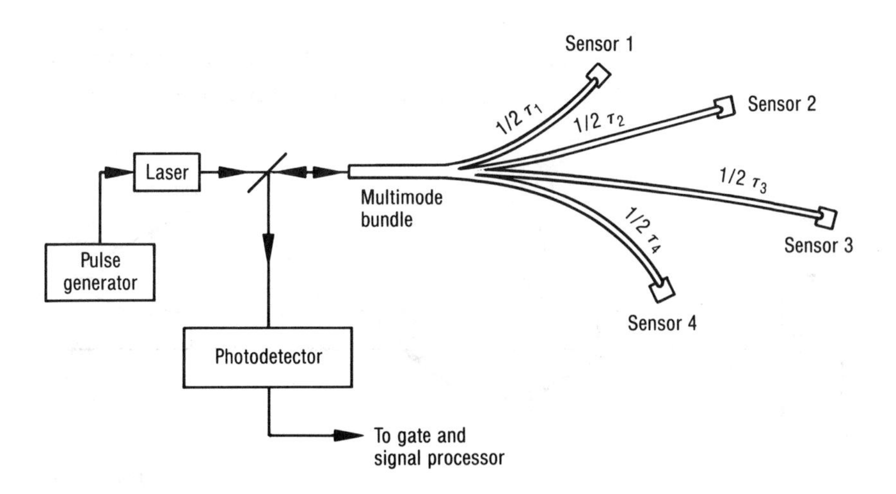

Figure 14-9
Star Network

delay for each sensor. Figure 14-10 illustrates a "T"-coupled sensor used in conjunction with a star coupler. Such configurations can result in hundreds of sensor points in a single system.(1)

Figure 14-11 shows a parallel network in which the transmitting fiber is separated from the receiving fiber.(2,6) The sensors are located on cross arms. The advantages of this approach are numerous. First, if a sensor fails, the whole network is not affected. Second, the modulated range, i.e., the change in intensity at each sensing point, can be made greater than for in-line sensors. This approach allows any sensor type to be used. Therefore, intensity-modulated and phase-modulated sensors could be used in the same network. The major disadvantage is that a large number of couplers are required.

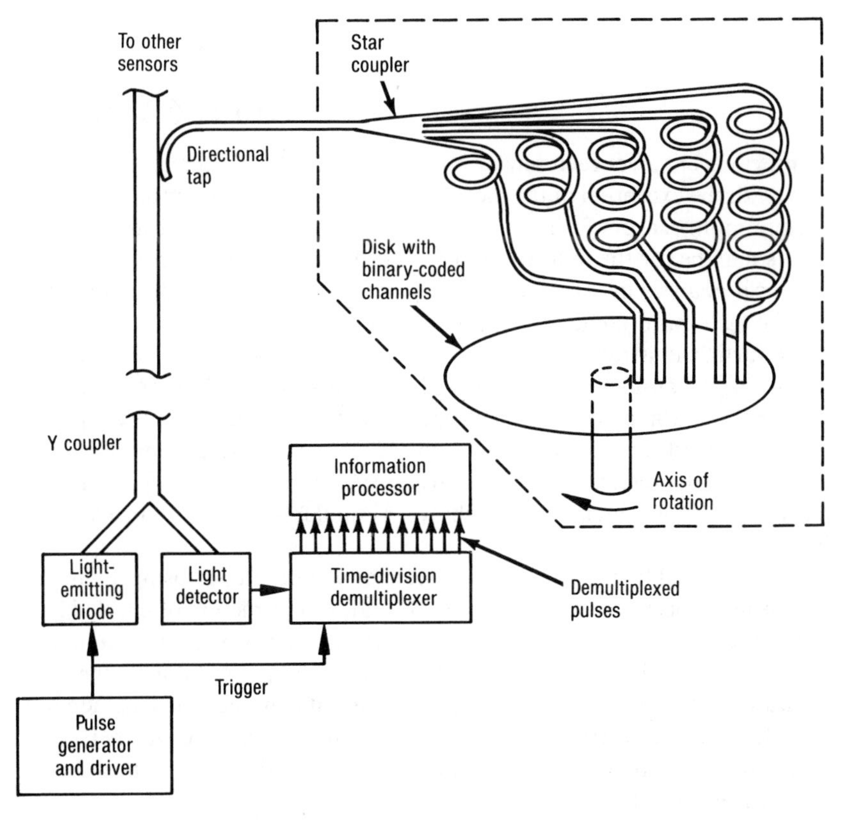

Figure 14-10
"T" Coupled/Star Network

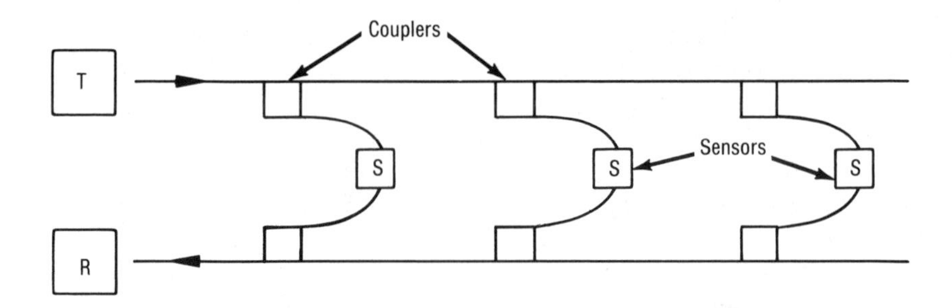

Figure 14-11
Parallel Network

(reprinted by permission from SPIE)

APPLICATIONS

In previous chapters each sensor has been treated as a discrete unit. However, with the exception of the most simple applications, nearly all situations require multi-sensors to input a central control area. The sensing requirements may be for the indication of the need for damage control on a ship, chemical analysis in a petrochemical plant, or robotic control in an automobile factory, to name just a few.

To further illustrate multi-sensors and sensing networks, some examples are given.(4) Consider monitoring a pipeline as shown in Figure 14-12. The concept uses a microbending concept. Any disturbance along the pipeline will microbend the fiber and cause an OTDR signal, which can detect the position of the disturbance. The sensing system, in this case, provides an alarm condition. Figure 14-13 shows a similar concept to provide an intruder alarm.

A distributed fiber optic hot spot detector was developed using an intrinsic sensing concept.(7) At the hot spot, the index of refraction of the cladding material changes due to temperature, and the transmission level is altered as shown in Figure 14-14. Using an OTDR approach in a continuous fiber, the position of the hot spot can be identified. Specific applications include hot spot detection in electrical power cables and high voltage transformers. In a straightforward manner this concept can be extended to an area heat detector for fire protection.

In summary, as fiber optic technology expands, the logical extension of the technology is toward sensing networks. Intrinsic sensors will provide the basis for low cost (per point) sensing systems.

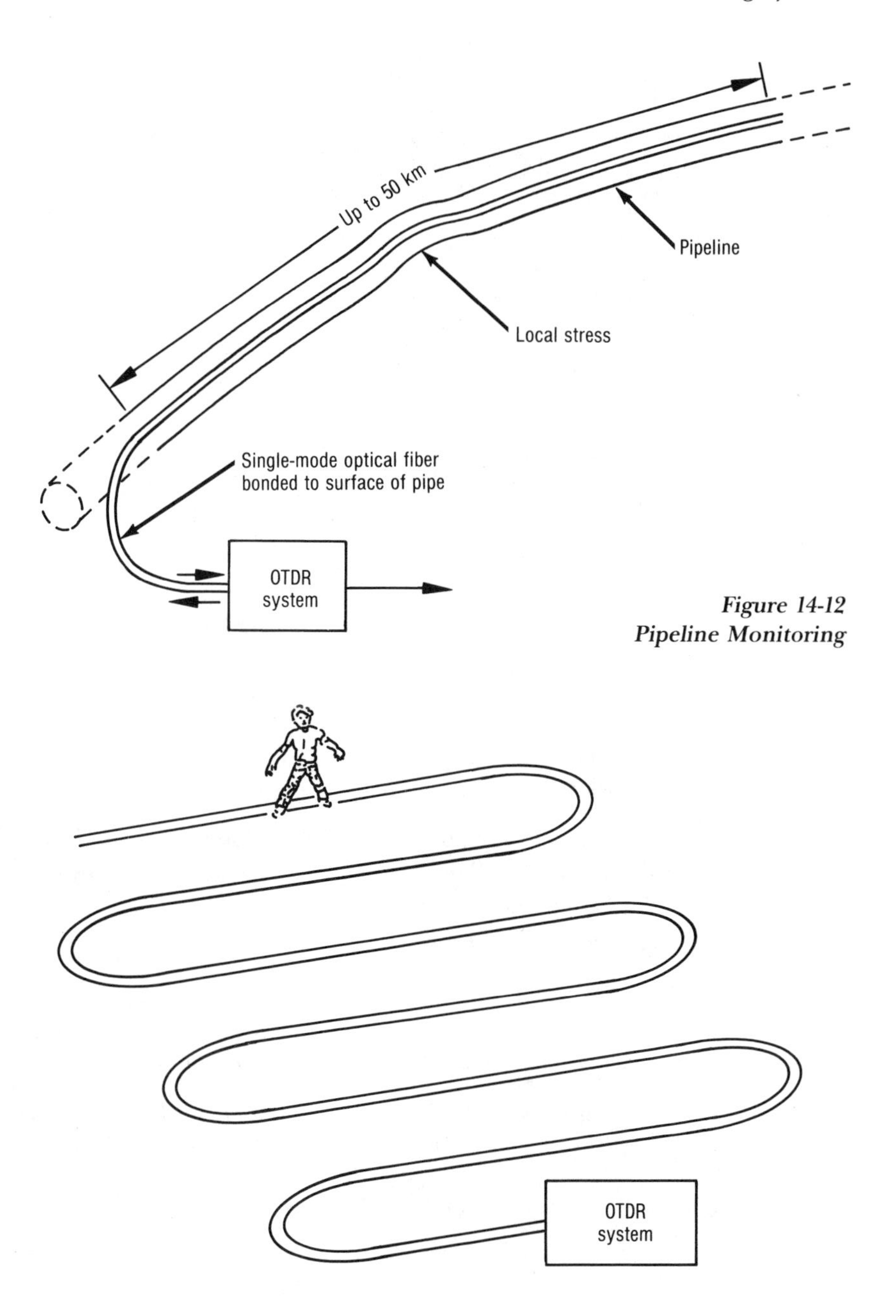

Figure 14-12
Pipeline Monitoring

Figure 14-13
Intruder Alarm (Security Fence)

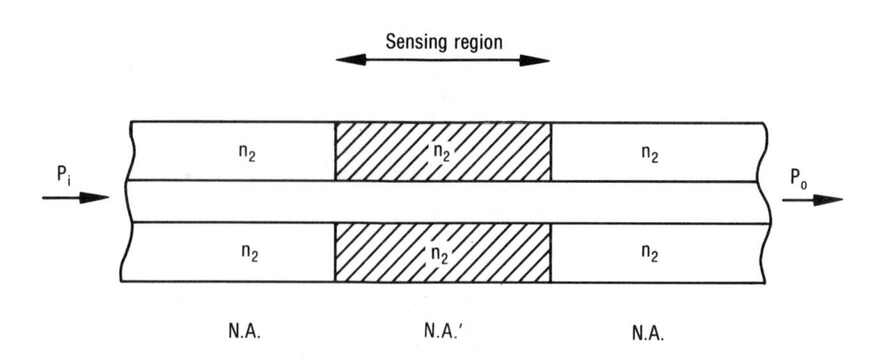

Figure 14-14
Intrinsic Fiber Optic Sensor Using Refractive Index Change in a Continuous Single Fiber Hot Spot Detection System
(reprinted by permission from SPIE)

REFERENCES

1. McMahon, D., Nelson, A., and Spillman, W., Jr., December 1981, "Fiber Optic Transducers," *IEEE Spectrum*, pp 24–29.

2. Davies, D., 1985 "Optical Fibre Distributed Sensors and Sensor Networks," *SPIE-Fiber Optic Sensors*, Vol. 586, pp 52–57.

3. Krohn, D., 1984, "Chemical Analysis with Fiber Optics," *Instrumentation and Control in the 80's, Proceedings of the 1984 Joint Symposium*, Des Plaines, Ill., pp 43–50.

4. Kingsley, S., 1984, "Distributed Fiber Optic Sensors," *Proceedings of the ISA*, Houston, TX, pp 315–330.

5. Kingsley, S., 1986, "Advances in Distributed FODAR (Fiber Optic Detection and Ranging)," *SPIE 30th Technical Symposium*, Vol. 718.

6. Davis, C., Carmone, E., Weik, M., Ezekiel, S., and Einzig, R., 1982, *Fiber Optic Sensor Technology Handbook*, Dynamic Systems, Reston, Virginia.

7. Kingsley, S., and McGinniss, V., 1986, "Distributed Fiber Optic Hot Spot Sensors," *SPIE 30th Technical Symposium*, Vol. 718.

Index

About the Author

Dr. David A. Krohn was the founder and President of EOTec Corporation until its recent sale to 3M. He is now the General Manager of the EOTec/3M subsidiary. The company focus is in the area of fiber optic sensors and communications for factory automation. He has done extensive research on intensity modulated fiber optic sensors.

Dr. Krohn has authored more than 35 technical papers and holds 23 patents all related to glass and fiber optic technology. He is a past Director of the Electro-optics Division of the ISA.